Tsunami Vulnerability Studies for the Coastal Regions

This book covers the deformation formulation at the source, the initial tsunami wave, its propagation, and run-up, along with the mathematical formulation for generating the initial wave, its propagation, run-up, and inundation. It discusses the varying nature of water depth in different regions, tsunami arrival times, and wave heights on the coast, including the role of islands near the coast acting as waveguides. Tsunami-induced forces on structures are addressed, including both non-breaking and breaking wave regions, as well as tsunami force and pressure.

- Provides a multi-perspective discussion of tsunamigenic phenomena affecting coastal regions
- Includes studies of tsunamis generated by meteoritic impacts and landslides and their effects
- Offers a complete overview of both mathematical and observational aspects of the tsunami phenomenon
- Explores comparisons between numerical results and on-site tsunami run-up and inundation observations
- Reviews all aspects of tsunami generation, wave propagation, and inundation at coastal sites

This book is aimed at researchers and professionals in geology, geophysics, seismology, natural hazards, and earthquake engineering.

Tsunami Vulnerability Studies for the Coastal Regions

Kirti Srivastava, Mounica Jakkula, and V.P. Dimri

CRC Press
Taylor & Francis Group
Boca Raton London New York

CRC Press is an imprint of the
Taylor & Francis Group, an **informa** business

Designed cover image: © Kirti Srivastava, Mounica Jakkula, and V.P. Dimri

First edition published 2026
by CRC Press

2385 NW Executive Center Drive, Suite 320, Boca Raton FL 33431
and by CRC Press

4 Park Square, Milton Park, Abingdon, Oxon, OX14 4RN

CRC Press is an imprint of Taylor & Francis Group, LLC

ISBN: 9781032553962 (hbk)
ISBN: 9781041327899 (pbk)
ISBN: 9781003784500 (ebk)

DOI: 10.1201/9781003784500

Typeset in Times
by Newgen Publishing UK

Dedication

We dedicate the book to our parents, who gave us the gift of dreams and the ability to realize them, and who are our role models in guiding us through our journey.

Contents

Foreword

The vast coastline of India, stretching over 7,500 kilometers, is a cradle of rich cultural heritage and ecological diversity, yet it remains inherently vulnerable to a range of natural hazards, among which tsunamis are particularly catastrophic. Tsunamis, characterized by their rapid onset, high energy release, and potential for widespread coastal inundation, pose a persistent threat to human settlements, infrastructure, and ecosystems along India's coastal regions. A comprehensive understanding of the mechanisms governing tsunami genesis, propagation, and coastal impact, alongside detailed vulnerability assessments of India's coastal regions, is imperative to inform mitigation strategies and enhance resilience. The authors have meticulously explored the geophysical mechanisms that trigger tsunamis, including their modeling from generation through propagation, run-up, and inundation. Special attention is given to the aftermath of the devastating 2004 Indian Ocean tsunami, which reshaped our approach to tsunami science and disaster management.

The book also discusses other facets of tsunami modeling, such as the forces exerted on structures, and tsunamis caused by landslides and meteorite impacts. Identification of high-risk areas, taking into account diverse bathymetric features and vulnerability mapping, provides critical insights to guide land-use planning and the design of protective infrastructure. The authors advocate for the implementation of ecosystem-based mitigation strategies, highlighting the protective functions of mangrove forests and coral reef systems as natural buffers that attenuate tsunami wave energy.

This book will prove invaluable not only to the scientific and technical communities but also to decision-makers seeking effective solutions to this critical national challenge. This work is a critical resource for multidisciplinary stakeholders, including geoscientists, coastal engineers, urban planners, and policy makers engaged in disaster risk reduction. Public awareness and community engagement are highlighted as vital components of effective disaster management, ensuring that local populations are better prepared and resilient when faced with potential tsunami threats.

I welcome the authors on this important journey and commend their contribution to strengthening our understanding and preparedness against one of nature's most formidable hazards.

Dr Prakash Kumar
Director
CSIR-National Geophysical Research Institute
Uppal Road, Hyderabad-500007, India

Preface

The havoc created by some tsunamis is still fresh in the minds of humankind, and it has alerted the entire scientific community to think about the tsunami warning system. Even if people are evacuated on time, the coastal economy can still suffer a big blow. Chapter 1 discusses the causes of tsunamis and historical tsunamis around the world.

Chapter 2 deals with the deformation formulation at the source, the initial tsunami wave, its propagation, and run-up. After a tsunamigenic earthquake, a permanent uplift and subsidence of the seafloor pushes the entire water column up and down. The potential energy from the water upliftment above mean sea level is then dissipated in all directions, horizontally propagating as a tsunami wave. To calculate the tsunami, the initial wave due to the different source mechanisms, such as earthquake, landslide, and meteorite, is first generated and is mandatory since the wave created by the movement of the seafloor has to travel forward in a highly incompressible fluid. In this chapter the earthquake source is considered and the fault parameters play an important role in calculating the initial wave generation. The chapter will discuss the mathematical formulation for wave propagation, run-up, and inundation in detail.

Chapter 3 discusses the varying nature of the depth of water in different regions due to the changing width of the shelf, which is responsible for the variations in tsunami arrival times and wave height on the coast. The depth of the water column varies with the width of the shelf, so tsunami arrival times are different and, to a great extent, dependent on the shelf width. Examples of the tsunami's reach at various locations shall be discussed. Also, the role of islands near the coast in acting as waveguides is discussed. The chapter also discusses the effect of ridges on tsunami propagation and arrival times at various locations. In the presence as well as with the absence of Lakshadweep Ridge's bathymetry, tsunami along India's south west coast is studied considering the 1945 Makran tsunami at the Subduction Zone of Makran. For India's southwestern coastal region the Islands of Lakshadweep acted as a wave barrier.

Several researchers have carried out tsunami-induced forces on structures. These studies have been incorporated into engineering mega structures like bridges and dams. But in the case of small buildings and residential houses, no such studies have been done, and there are no guidelines to ensure their protection during a tsunami. Chapter 4 attempts to calculate the maximum forces along the coast for various tsunami events. Calculating tsunami-induced forces along the coast is first categorized into different regions based on the maximum wave heights, water depths and slope. Then, for the non-breaking wave region, the breaking wave region, and the broken wave region, the tsunami force/pressure are calculated separately. Examples from various areas that have suffered damage and were impacted by tsunamis are studied. A sea wall that is the same height as boulders is considered to estimate the inundation at the location and also the surge forces along the sea wall are calculated. Also, we calculate the pressure/force of the non-breaking, breaking, and broken wave regions separately.

Chapter 5 discusses the creation of a tsunami by meteorite impact in the Indian Ocean. The 1950 DA asteroid, forecasted to fall on Earth on 16 March 2880, is being studied. Though the likelihood of this event is very low, this research helps us understand beyond the limitations of earthquake-generated tsunamis. Using examples, the effect of a meteorite impact on the coast is discussed.

In Chapter 6 yet another aspect of the tsunami due to a landslide is presented. First, we will identify petroliferous basins along the continental margin, such as those with steep topographic gradients, high sedimentation rates, and fluid/gas movement that causes slumping or sliding of upper slope sediments. This can result in a landslide tsunami and one can quantify the hazard associated with such a landslide tsunami. The mathematical formulation is discussed in detail. Finally in Chapter 7, the results, along with the limitations and scope for future work, are summarized.

Finally, we thank our family and friends who stood by us and gave us the confidence to achieve our dreams and goals. The help rendered by my students Ms Aruna Kumari, Dr Vikas and Dr Farveen Begum during the book writing is incredibly appreciated. My special thanks go to Dr (Late) Swaroopa Rani, Dr Swapna, Dr Krishna Kumar, and Dr Shila for their contributions to tsunami studies. The authors sincerely acknowledge Dr Prakash Kumar, Director CSIR-National Geophysical Research Institute, Hyderabad, India, for his constant encouragement.

About the Authors

Kirti Srivastava received a PhD (Geophysics) from Osmania University, in 1991 under the guidance of Dr V.P. Dimri, CSIR-NGRI, Hyderabad, India and a Post Doctoral from the Dept. of Civil and Environmental Engineering, Temple University USA on Stochastic Differential Equations. She is a retired Chief Scientist and was the Head of Earth Process Modeling Group at CSIR-National Geophysical Research Institute up to January 2022 and later Raja Ramanna Fellow of DAE (Department of Atomic Energy) at CSIR-NGRI. Presently she is WIDUSHI Fellow of DST Govt of India at CSIR-NGRI. Kirti Srivastava's research areas include the Mathematical and Stochastic Modeling of non-linear behavior of Earth systems. She has carried out Tsunami hazard studies for the Kalpasar dam being proposed along the Gulf of Khambhat by the Gujarat Government. An impact study was carried out for tsunami-induced hydrodynamic forces on the proposed dam structure. She has carried out tsunami hazard studies due to earthquakes, landslides and meteorite impact along the Indian coast. She has also carried out tsunami hazard modeling at nuclear installations in India (for the Atomic Energy Commission, Bhabha Atomic Research Centre, BARC). Her basic research includes the quantification of accurate subsurface temperature, heat flow and rheological field, the thermal state of the crust in seismically active regions, lithospheric thicknesses from the thermal point of view for different terrains in the Indian shield, and the modeling of 2D thermal state across seismic profiles in the India region. In addition she has worked on stochastic nonlinear transient groundwater flow modeling for one and two-dimensional flow with uncertainties in sources and parameters with applications to quantify the state of the aquifers. Her PhD work is related to the development and application of seismic deconvolution techniques to offshore seismic data with a special emphasis on time-varying deconvolution for the removal of noise, and a multichannel adaptive deconvolution technique developed for the removal of water bottom multiples keeping in view the stability and accuracy of the solution. She has several National and International publications, and two US Patents to her credit. She has guided seven PhD students, three of whom are pursuing their PhD under her guidance. She is the recipient of the National Merit Scholarship in High School, the CSIR Young Scientist Award in 1995, and is a Fellow of the AP Academy of Sciences and Fellow of IGU. She was the Senate-Member of the National Institute of Technology (NIT-Warangal) ,Member of the Management Council & Academic Council of SRTM University Nanded, Maharashtra, a subject expert for the Indo-US Fellowship for Women of DST, and is presently a member of the Women Scientist Scheme, DST, Govt of India. She has served in different capacities in Vasavi Engineering College, Gokaraju Rangaraju Engineering college, and Arora Colleges.

Mounica Jakkula received her PhD from AcSIR (Academy of Scientific and Innovative Research), CSIR–National Geophysical Research Institute (NGRI), Hyderabad, in 2021, under the guidance of Dr. Kirti Srivastava. She holds an Integrated MSc (5-year) in Mathematics and Scientific Computing from IIT Kanpur. Mounica Jakkula's research areas include numerical and mathematical modeling of tsunami generation and propagation due to earthquakes and meteorite impacts. She has extensive experience with the TUNAMI-N2 code and has developed FORTRAN and MATLAB models for initial wave generation (meteorite impact), and estimation of tsunami-induced forces on coastal and offshore structures. She has contributed to tsunami hazard assessment studies for the Indian coastline, including the Kalpasar Project. She has published her research in national and international journals and has presented her work at several scientific conferences. She won the best paper award at the 19th NSSS conducted by ISRO for her work on meteorite impact tsunamis. She is currently based in Munich, Germany. Using the Adomian method of decomposition she has solved different kinds of equations like parabolic and hyperbolic non-linear, and undertook a project under Dr Kirti Srivastava for her MSC dissertation work on heat equation problems for the quantification of subsurface thermal structures. Her hobbies include badminton, dance, fine arts, high altitude trekking and motorbiking

V.P. Dimri is the former President of the 36th International Geological Congress appointed by the Union Cabinet and Prime Minister, Shri Narendra Modi in 2017. He was a Distinguished Scientist, at the CSIR-NGRI, Hyderabad, India for three years (2010–2013) and INSA Senior Scientist for three years (2013–2016). He was Dr. Homi Bhabha Chair Professor at Central University of Hyderabad from Nov 2016 until he was appointed President, the 36th IGC. He served as the Director of the NGRI from October 2001 to February 2010. Also, he served as the Director, Gujarat Energy Research and Management Institute, Gandhinagar, Gujarat, India. He has more than 50 years of experience and has published 135 research publications in refereed journals. He has authored and edited books by several internationally reputed publishing houses. He has produced nine PhDs and guided many postdoctoral researchers. He was the project leader of many national and international projects and has served as a member and chairman of many national and international committees. He was conferred with the highest civilian award "Padma Shri" by the Govt. of India in 2010. He delivered prestigious "Sir Axford Lecture" to the Asia Oceania Geosciences Society, Singapore; in the same year the other Sir Axford lecture was delivered by the Nobel laureate Prof. Lee of Taiwan. He is first Asian to receive the Lorenz Award Lecture by the American Geophysical Union in the field of non-linear geophysics. Among the many other important awards conferred on him are Chatterjee Award by the Indian Science Congress, Outstanding Scientist Award by FAPCCI, Department of Ocean Development Award, National Mineral Award and National Geoscience Award by MoES, S&T Excellence Award DST, Uttarakhand. He delivered the Prof. K. Naha memorial medal lecture in 2016, awarded by INSA, New Delhi. The R.C. Mishra Lifetime Achievement award was conferred to him in 2019. He is a Fellow of many learned academies and societies among which are Third World Academy of Sciences, Indian National Science Academy, National Academy of Sciences, Andhra Pradesh Akademi of Sciences, Telangana Academy of Sciences, Geological Society of India, and Indian Society of Applied Geochemists. He was past President of the Andhra Pradesh Akedemi of Sciences, Indian Geophysical Union. Currently, he is an Honorary Scientist of INSA at CSIR-NGRI and President of the Federation of Indian Geoscience Association. He stands among the top 100 great Indians dedicated to the service of our nation.

1 Introduction

1.1 INTRODUCTION TO TSUNAMI

A tsunami is a series of gravity waves of extremely long wavelengths and periods, usually generated by undersea earthquakes, volcanic eruptions, landslides, or meteorite impacts. In the past, seismic ocean waves were inaccurately referred to as "tidal" waves, erroneously suggesting a direct association with the tides. The Japanese term "tsunami" was introduced "tsu," meaning "harbor," and "Nami," signifying "wave", emphasizing their seismic origin and translating to "harbor wave." Tsunami waves exhibit remarkable characteristics, traversing vast distances and spanning entire ocean basins while preserving minimal energy loss. Propagating at speeds exceeding 800 km/h, they change as they approach shallow waters. Tsunami waves steepen rapidly, leading to a sudden increase in height. This phenomenon can result in the inundation of low-lying areas, and in some cases, waves may break, causing extensive damage (Dawson and Stewart, 2007; Marghany, 2014).

They can cover vast distances and traverse the entire ocean basins with minimal energy loss and propagate as standard gravity waves with a typical period ranging from 5 to 60 minutes. As tsunami waves approach the shallow waters, they steepen and rise in height, thus flooding low-lying areas and potentially causing significant destruction. When these waves travel from deep waters to shallow waters they slow down at the wave base yet their period remains unchanged while their height increases. When large volumes of seafloor elevate or subside, a tsunami is generated.

November 5 was designated as the World Tsunami Awareness Day by the UN General Assembly. Siegel (2020) has said that it is anticipated that coastal regions like Asia, Africa, and to some extent, South America will see substantial population growth in the coming years. Despite their rarity tsunamis are one of nature's deadliest natural calamities. Over several lakhs of people have died as a result of tsunamis in the last century. The deadliest of these was the 26 December 2004 Indian Ocean tsunami which killed lakhs of people. The most severely affected nations were Indonesia, Sri Lanka, India, and Thailand. The rapid urban development and rising tourism in regions vulnerable to tsunamis are placing increasing numbers of people in danger. This suggests that the hazard level may be alarmingly high. Hence, it is crucial to implement prompt and coordinated measures to mitigate the tsunami risk. Past tsunami occurrences, including the symbols that represent the causes and the deaths from earthquakes, volcanic eruptions, and landslides, are shown in Figure 1.1 (https://ngdc.noaa.gov/).

The release of substantial energy is crucial for deforming the ocean bottom and uplifting the seafloor to significant heights. They are generated when extensive volumes of seabed either rise or collapse, signifying a seismic event of large magnitude (Sun, 2020).

To mitigate the risks associated with tsunamis, it is essential to invest in early warning systems and public education. There is a need for vulnerability mapping, inundation modeling, run-up

DOI: 10.1201/9781003784500-1

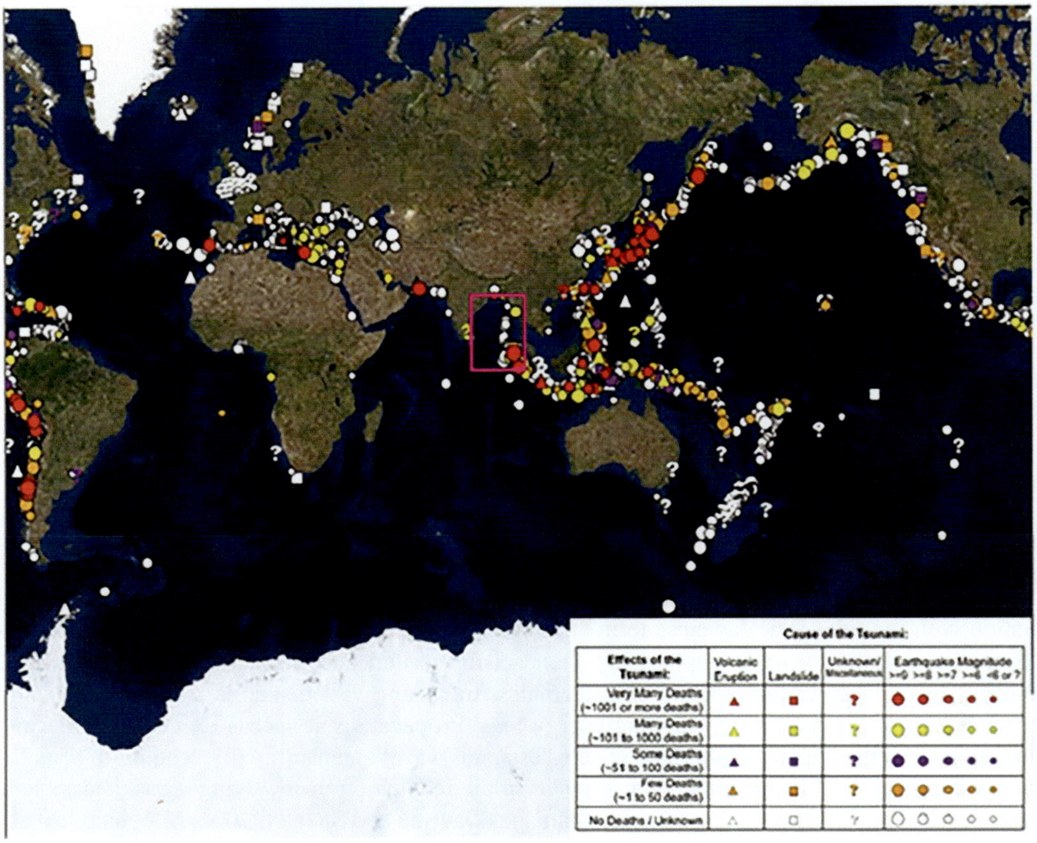

FIGURE 1.1 Tsunami events, incorporating symbols to indicate causes and fatalities for earthquakes, volcanic eruptions, and landslides.

and arrival time analysis, and shoreline change monitoring mapping. Vulnerability mapping helps identify and prioritize high-risk areas, facilitating targeted disaster preparedness and response. Inundation modeling and run-up analysis predict the extent and impact of tsunami waves and help in effective evacuation planning and infrastructure design. Analyzing arrival times allows for early warning systems, giving populations more time to evacuate. Shoreline change monitoring assesses the impact of natural and anthropogenic factors on coastal resilience. Together, these tools provide a comprehensive approach to reducing tsunami risk and enhancing coastal resilience.

The data regarding tsunamigenic events across various oceans indicates that 76% are in the Pacific, 10% in the Mediterranean, 9% in the Atlantic, and only 5% in the Indian Ocean. This demonstrates that most tsunamis are concentrated in the Pacific Ocean. (Source: http://tsun.sscc.ru/tgi_5.htm).

1.2 PHYSICAL CHARACTERISTICS OF TSUNAMI

Tsunami waves are characterized by their ability to cover vast distances with minimal energy loss; they can steepen and amplify in height as they near shallow waters, leading to potential flooding and significant damage in low-lying regions. Tsunamis range in size from centimeters to more than 30 m in height. In deep water, tsunamis are hardly more than 1 m tall and will not be recognized by ships because of their long periods. Tsunami heights can vary greatly along the coastal region. Not all earthquakes generate tsunamis as enormous amounts of energy need to be released during an earthquake to deform the ocean bottom and to uplift the seafloor to significant heights. When large volumes of seafloor elevate or subside, a tsunami is generated. Only thrust earthquakes are known to generate tsunamis with magnitudes greater than 7.

Waves generated by the wind typically exhibit a period ranging from five to twenty seconds and a wavelength between 100 to 200 meters. In contrast a the period of a tsunami can be from 10 minutes to two hours with a wave length varying from 10 to 500 km. This unique characteristics of a tsunami distinguishes them from swell or storm waves. A shallow water wave is when the ratio of the water depth (d) to its wavelength (L) becomes very small ($d/L < 0.05$) Given that ocean depth is relatively shallow compared to the wavelength of the tsunami it is generally characterized as a shallow water wave.

The height of the tsunami waves can be estimated using linear theory as they traverse the ocean, experiencing shoaling and refraction. The energy loss rate of the wave is inversely proportional to its wavelength. Given that a tsunami possesses an exceptionally large wavelength, it retains most of the energy during propagation Bryant and Nott (2001). Hence, in deep water, a tsunami will move at a large speed with minimal energy loss.

As tsunami waves move from deep waters to shallow waters they experience significant transformation. This phenomenon occurs because the group velocity, which represents the speed at which the wave energy is transported, reduces as the water depth decreases. To sustain a consistent energy flux under stationary conditions, this reduction in transport speed necessitates an increase in energy density. This process is referred to as the shoaling effect. Consequently, there is a decrease in wavelength, and the height of the tsunami increases. Near the shoreline, this phenomenon is most accurately characterized by solitary waves. A solitary wave retains its shape as it approaches shallow waters, and due to the uniform distribution of tsunami kinetic energy throughout the water column, minimal energy is lost, particularly along steep coastlines.

1.3 CAUSES OF TSUNAMI

Tectonic activity (earthquakes) is the major contributor; 85% of global tsunamis result from underwater earthquakes, as highlighted by Whitmore et al. (2008). Landslides cause about 8%, 5% are due to volcanic activity, and less than 2% arise from a meteorite impact.

1.3.1 Submarine Earthquake

Large to great earthquakes are common at the subduction zones, where the heavier oceanic plates subduct beneath the lighter continental plates. During the inter seismic period, the subducting plate experiences stress accumulation, which is eventually released in the form of fault slips during an earthquake. If the seismic event takes place underwater, it has the potential to generate a tsunami (Figure 1.2). A submarine earthquake can only generate a tsunami if the magnitude is greater than 7 on the Richter scale and focal mechanism should be thrust or normal type and a shallow focal depth. The circum-Pacific subduction zone is the most common tsunamigenic region followed by the Indian Ocean, which forms an important tsunami region with Andaman-Sumatra and Makran subduction zones as the primary earthquake source regions. Some of the great tsunamigenic earthquakes around the world are listed in Table 1.1.

The historical tsunami listed in Table 1.1 show that the majority of tsunamis are generated along subduction zones and the major plate boundaries in the Pacific, the Atlantic, the Mediterranean, and the Indian Ocean regions. Several cases of small tsunamis are known to have originated along the mid-oceanic ridges and some major transform faults.

1.3.1.1 Lisbon Tsunami

The first breakthrough in modern seismology and earthquake engineering was the scientific study of Great Lisbon Earthquake of November 1, 1755 (Magnitude 8.5) which is known to have generated a

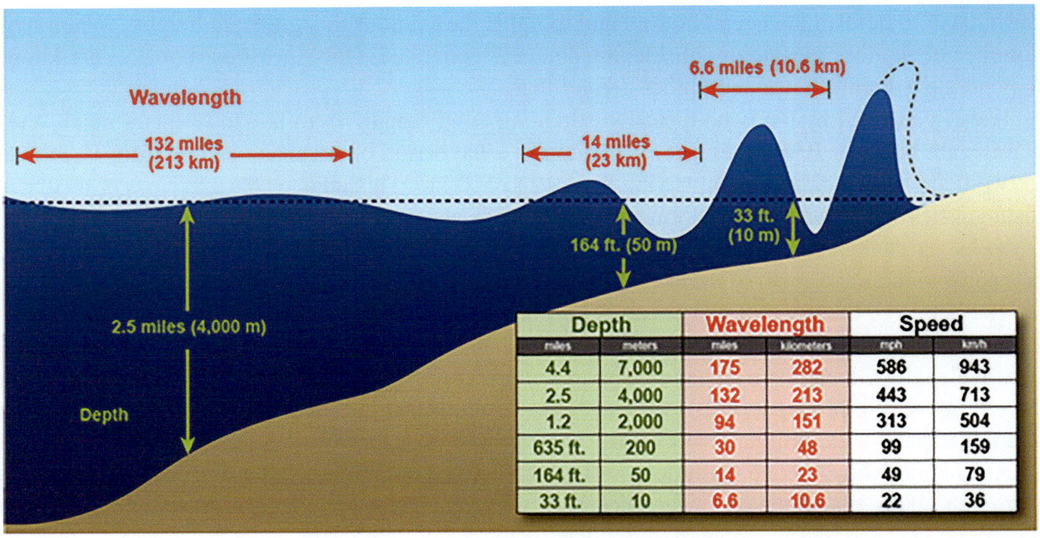

FIGURE 1.2 Tsunami waves approaching the coastline.

TABLE 1.1
Table of great tsunamis due to submarine earthquakes from across the globe

S.No	Date and location	Long. (DD)	Lat. (DD)	Mag M_w	Observations	Ref.
1	1.11.1755 Lisbon, Portugal			8.5	On Dec 1, 1755, an earthquake struck Lisbon, Portugal, resulting in a tsunami that caused severe damage to Portugal, Spain, North Africa, and the Caribbean	Abe (1989)
3	4.11.1952 Kamchatka, Russia			9.0	The property damage from these waves is estimated to be around millions of dollars	Kanamori (1977)
2	22.05.1960 Valdivia, Chile	73.4	38.14	9.5	One of the most powerful event that triggered a devastating tsunami of the 20th century the tsunami wave heights in some locations were around 25 m	Fujii et al. (2013)
3	28.03.1964 Alaska, USA	-147.7	61	9.2-9.3	The Alaska tsunami wave heights reached 67 m	Ichinose et al. (2007)
4	26.12.2004 Sumatra Indian Ocean	94.57	3.83	9.3	This earthquake triggered a series of tsunami waves that ravaged the surrounding coastal areas	Gupta (2005)
5	11.03.2011 Tohoku	142.3	38.3	9.1	Tsunami wave heights reached upto 40 m	Mori et al. (2011)
6	29.07.2025 Kamchatka, Russia	160.2	52.53	8.8	Tsunami alert was issued across the Pacific region and at Shumshu it was around 19 m	USGS

transoceanic tsunami. Abe (1979, 1989) obtained a magnitude scale based on the tsunami amplitudes obtained at the tide gauges and from this scale they obtained the seismic moment estimate. The city of Lisbon, one of the main harbors in Europe during the 17th and 18th centuries, was severely damaged by this tsunami (Baptista et al., 1998). It is known to have caused severe destruction along the coastal areas of Portugal, Spain, and North Africa, although there is no actual estimate on the loss of life. Gupta and Gahalaut (2013) describe that the tsunami caused heavy damage to ships and coastal towns along the west coast of Portugal.

1.3.1.2 Kamchatka Earthquake, November 4, 1952 (Magnitude 9.0)

On 4 November 1952 a mega-thrust earthquake of Magnitude 9.0 occurred at a depth of 22 km which generated a Pacific-wide tsunami that struck the Hawaiian Islands. Here the Pacific plate is subducting below the Okhotsk micro plate. Because of this subduction process the Kuril island chains, volcanoes along the entire arc and a deep Kuril-Kamchatka trench have been formed. The November 4, 1952 Kamchatka earthquake generated a tsunami and the property damage from these waves is estimated to have been millions of dollars. Calebaugh (1977) reported much damage on Oahu's north coast as the wave heights were around 9.1 m at Kaena Point and Waialua. The tsunami caused the loss of life on the Kamchatka peninsula and northern Kuril islands. The tsunami waves were also observed in Alaska. At Massacre Bay, Attu, the wave height was around 2.7 m, and on the east coast of Kamchatka the tsunami waves were up to 13 m high. Kanamori (1977) showed that in California the wave height ranged from 1.4 m at Avila to 1.0 m at Crescent City.

1.3.1.3 Chile Earthquake, May 22, 1960 (Magnitude 9.5)

The Valdivia earthquake was one of the largest mega-thrust earthquake of the 20th century with a rupture length of 1000 km. It generated a very destructive tsunami. Chile is located in a highly seismically active regions of the world, along the Pacific Ring of Fir. This tsunamigenic earthquake was a mega thrust earthquake of magnitude 9.5 (Fujii et al., 2013). The shallow earthquake was at a depth of 33 km. Several thousand people were killed during this tsunami, and it caused large-scale devastation along the coast of Chile from Lebu to Puerto Aisen (Kanamori, 1977). Several ships were battered by the currents and the tsunami waves destroying the coastal areas. The earthquake's epicenter was about 570 km south of Lumaco. The tsunami destroyed the city of Valdivia with waves up to 25 m high. This global tsunami affected the southern parts of Chile, Hawaii, Japan, the Philippines, eastern parts of New Zealand, and southeast Australia (Van Dorn, 1987).

1.3.1.4 Alaskan Earthquake, March 28, 1964 (Magnitude 9.3)

The Alaskan earthquake on March 28, 1964 took place where the Pacific and North American plate ruptured near Prince Williams Sound. The depth of the earthquake was about 25 km. This great earthquake and its subsequent tsunami are known to have taken the lives of several people and caused the loss of property worth millions of dollars. Severe property damage took place at Anchorage, about 120 km northwest of the epicenter (Ichinose et al., 2007). This magnitude 9.3 mega thrust earthquake remains the most powerful earthquake recorded in North America. This earthquake triggered massive landslides and rockslides which generated large waves up to 70 meters high. Major structural damage due to liquefaction and landslides caused severe damage to buildings and infrastructure. This tsunami traveled to Hawaii and Japan causing severe damage. Plafker et al. (1969) described how the submarine landslide triggered tsunami in several fjords in Prince Williams Sound.

1.3.1.5 Sumatra Earthquake, December 26, 2004 (Magnitude 9.1–9.3)

The Sumatra earthquake of 26 December 2004 measuring 9.1 ranks as the third most powerful earthquake measured globally since 1900 (Gupta, 2005). This catastrophic mega thrust earthquake took place in a seismically active area where the Indian plate is being subducted beneath the Burmese

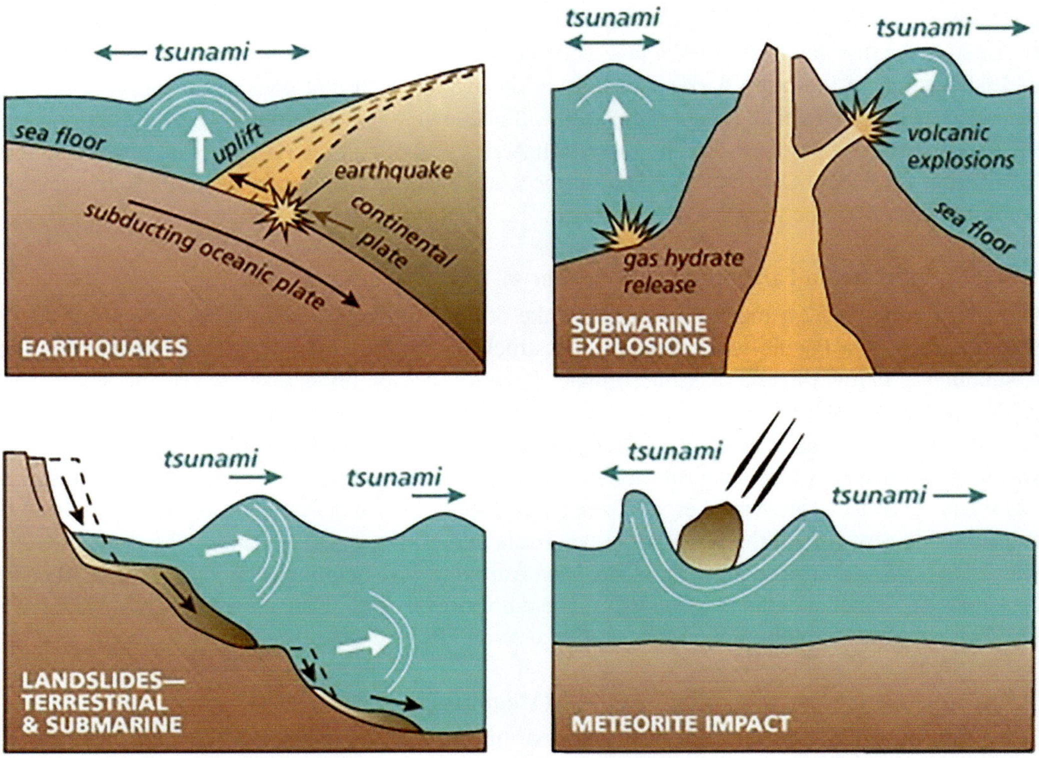

FIGURE 1.3 Generation of a tsunami due to different sources.

Source: www.noaa.gov/media/image_download

plate at the Sunda trench. The Andaman Sumatra subduction zone represents one of the world's most active plate boundaries facilitating an oblique northward convergence of over 50 m per year between the south Asian and Indo–Australian plates (Stein and Okal, 2005). The earthquake originated in the Andaman-Sumatra Subduction Zone (ASSZ), rupturing 1200 km, and with a focal depth of 25 km. It triggered a trans oceanic tsunami that caused extensive damage in the Andaman region and along the eastern coast of India. This tsunami impacted numerous nations from Indonesia to Somalia resulting in hundreds of thousands of people being killed or missing and displacing millions of people across 14 countries in South Asia and East Africa (Park et al., 2005). It is estimated the total death toll was 227,898 (U.S. Geological Survey). After the 2004 Sumatra tsunami, India established the Indian Tsunami Early Warning System (ITEWS) at Indian National Centre for Ocean Information Services (INCOIS), to study and provide tsunami hazards and warnings to coastal communities in India (Kumar et al., 2012).

1.3.1.6 Tohoku Earthquake, March 11, 2011 (Magnitude 9.0)

On March 11, 2011, a magnitude 9.0 mega thrust earthquake struck near the northeast coast of Honsu, Japan, generating a massive tsunami. This event is among the most powerful earthquakes ever recorded in Japan. The Pacific plate is shifting westward at a rate of approximately 83 mm/yr relative to the North American plate and begins its westward descent beneath Japan at Japan Trench. Analysis of the earthquake rupture suggests that the fault had experienced an upward movement of 30–40 meters (Mori et al., 2011). The resulting tsunami due to this earthquake claimed thousands of lives and many individuals were reported missing. The earthquake's epicenter was located about 130 km east of Sendai with a focal depth of 30 km. The disaster caused extensive destruction to buildings, roads, bridges, and railway lines along the entire east coast of Honshu, from Chiba to Aomori. Most of the causalities and damage were reported in Iwate, Miyagi, and Fukushima due to this Pacific-wide tsunami which had a maximum run-up of 37.88 m, recorded at Miyako.

1.3.1.7 Kamchatka Earthquake, July 29, 2025 (Magnitude 8.8)

A magnitude 8.8 event on July 29, 2025 took place along the Kuril-Kamchatka trench. This is a large thrust fault region that stretches from the eastern shore of Kuril island to Kamchatka Peninsula. The tsunami waves were relatively small in most places but Russia declared a state of emergency in Shumshu, Kurli islands with the tsunami reaching 19 m. Japan also evacuated a several million of people near the Fukushima nuclear facility.

1.3.2 Submarine Landslides

The sea bed is characterised by peaks and troughs similar to those present on the Earth's surface. The unstable rock deposits, slopes, and debris can suddenly collapse in the water resulting in a landslide. These landslides can occur suddenly and cause the water column to rise forming the initial wave. The generation of the tsunami is influenced by the volume of the landslide material that displaces the water, the velocity of its movement, and the depth it moves to. In the case of an earthquake the sudden shaking may trigger a landslide that creates local waves that travel as tsunami waves to the coast causing large-scale destruction. The 1964 Alaska tsunami is the result of a magnitude 9.3 earthquake that triggered a rock fall which generated a tsunami causing widespread destruction, as documented by Landers (1996). Most earthquakes generate submarine landslides which can cause a landslide tsunami. The landslide tsunami generally loses energy quickly and rarely travels to distant coasts (Table 1.2).

TABLE 1.2
List of significant submarine landslide-generated tsunamis

Sr. No	Year	Location	Sliding volume & max. wave height	Observations	Reference
1.	10.07.1958	Lituya Bay, Alaska, USA	30.6 million m^3 and 524 m. Marked as the tallest tsunami in the world	On flat ground, the maximum inundation distance was 1400meters	González-Vida et al. (2019)
2.	18.11.1929	Grand Banks, Newfoundland, North America	200 km^3 and 13 m	12 telegraph cables were broken and 28 people were killed	Fine et al. (2005)
3.	09.10.1963	Vajont Reservoir, Casso, Italy	270 million m^3 and 235 m	The city of Langrone was Destroyed and 2000 people were killed	Panizzo et al. (2005)
4.	17.07.1998	Papua New Guinea	-	2,200 people were killed by the earthquake of magnitude 7.1, followed in 11 minutes by a tsunami of 15 metres due to a landslide which was triggered by the earthquake	Synolakis and Okal (2002); Tappin et al. (2008)
5.	17.10.2015	Taan Fiord, Alaska	180 million tons and 193m	The tsunami directly affected about 2 km^2 onshore while it inundated over 20 km^2	Higman et al. (2018)

1.3.2.1 Lituya Bay, Alaska

The Lituya Bay strike-slip earthquake of July 9, 1958 occurred on the Fairweather Fault, with a magnitude of 7.8 and 8.3. It triggered a rockfall that displaced 30 million cubic meters of material into the narrow inlet of Lituya Bay, Alaska. This sudden event generated a mega-tsunami of more than 524 meters at the Gilbert inlet. This rockfall removed about 400 m of ice from the Lituya Glacier and destroyed and eroded its rocky deltas. The waves generated by the rockfall had destroyed the vegetation and headlands along the bay region. On flat ground, the maximum inundation distance was 1400 meters González-Vida et al. (2019)

1.3.2.2 Grand Banks, Newfoundland, North America

The Grand Banks of Newfoundland consist of a collection of submerged plateaus located to the southeast of the Newfoundland Islands, situated on the North American continental shelf. On November 18, 1929, a major earthquake of magnitude 7.2 occurred about 400 km south of Newfoundland and triggered a submarine landslide. The tsunami generated waves between 3 and 13 m high along the Burin peninsula. Several residents along the southern coast of the peninsula were displaced. Fine et al. (2005) described how telegraph cables were broken and around 28 people were killed.

1.3.2.3 Vajont Reservoir, Casso, Italy

On October 9, 1963, a rock detached and plunged into the Vajont Dam, Italy, generating a colossal wave that displaced at least 50 million cubic meters of water. Although, the dam did not sustain any significant damage, the resulting flooding devastated several villages downstream of the dam, and many people lost their lives. Despite utilizing the most advanced technical knowledge, the dam was constructed without taking into account the geological assessment of the area. The city of Langrone was destroyed and 2000 people were killed (Panizzo et al., 2005).

1.3.2.4 Papua New Guinea

On July 17, 1998, at 5:49 pm local time, a 7.0 magnitude earthquake occurred in Papua New Guinea. This seismic event triggered a tsunami that reached a height of 15 m at Arop village. The tsunami was a result of a submarine landslide induced by an earthquake. The disaster resulted in the loss of 2,200 lives. Several researchers have done extensive work on this tsunami including Synolakis and Okal (2002) and Tappin et al. (2008).

1.3.2.5 Taan Fiord, Alaska

On October 17, 2015, a slope failure at the terminus of Tyndall Glacier resulted in a fall of 18 million tons of rock into Taan Fiord, Alaska. This event generated a tsunami that reached heights up to 193 meters, making it one of the highest tsunamis run-up recorded globally. Notably, there had been signs of deformation for decades before the failure, as clearly shown in sedimentary records. This evidence suggests that geological data can enhance our understanding of past similar events and may offer early warning signals. Higman et al. (2018) have shown that the tsunami directly affected about 2 km^2 onshore while it inundated over 20 km^2. Inverse and forward modeling can establish a comprehensive framework for understanding the geological implications and hazards associated with such events.

1.3.3 Submarine Volcanic Eruption

Catastrophic volcanic eruptions are sudden occurrences that displace a large volume of water and create extremely destructive tsunami waves in the immediate surrounding regions. This phenomenon arises from abrupt movement and the sudden displacement of water due to volcanic explosion, a volcano's slope failure, or, more frequently, a phreatomagmatic explosion coupled with the collapse or engulfment of the volcanic magmatic chambers. As an example consider the August 26,

1883, eruption and collapse of the Krakatoa (Krakatau) volcano in Indonesia. The waves generated reached astonishing heights of 135 feet leading to widespread destruction to coastal towns and villages along Sunda Strait in Java and Sumatra, resulting in the loss of 36,417 lives. Likewise, the fall of the Minoan civilization in Greece around 1490 BCE is linked to the explosion and collapse of the Santorin volcano in the Aegean Sea (Figure 1.2; Table 1.3).

1.3.3.1 Krakatoa, Indonesia, August 27, 1883

From May 20, 1883 to October 21, 1883 the volcanic islands of Krakatoa situated in Sunda Strait experienced a series of volcanic eruptions. The most catastrophic of these eruptions took place on August 27, 1883 resulting in the destruction of more than 70% of the island and its nearby archipelago as it erupted with such intensity that it collapsed into a caldera. Due to the eruption and the tsunami it created, 36,000 people were killed. Each explosion was accompanied by a tsunami estimated to be around 30 m high. It hit the port of Madras with 2 m run-up height (Bapat, 1982).

1.3.3.2 Ritter Island, Papua New Guinea, March 13, 1988

On Ritter Island a volcanic island located off the coast of Papua New Guinea, a large sector collapse took place on March 13, 1988, which generated a large tsunami. This is one of the most extensive lateral collapses of an island volcano in history, which produced tsunami waves that wreaked havoc on coastal communities throughout the Bismarck Archipelago and New Guinea, causing severe damage to life and property. The tsunami waves reached a height of 20 m in some islands, e.g., Sakar and New Britain. In New Guinea, hundreds of kilometers away, wave heights of up to 8 m were observed. This event serves as an important case study linked to volcanic sector collapses and its generated tsunami.

1.3.3.3 Anak Krakatoa, Indonesia, December 22, 2018

On December 22, 2018 Anak Karkatau experienced an eruption and a significant portion of the volcano's southwestern flank collapsed and plunged into the ocean, triggering a tsunami. The tsunami effected about 300 km of the coast of Sumatra and Java and a number of people along the coast were displaced. This is one of the deadliest volcanic event of the 21st century (Walter et al., 2019).

1.3.3.4 Hunga Tunga, January 15, 2022

The Hunga Tonga-Hunga Haapai volcano I Tonga erupted on January 15, 2022 releasing a substantial cloud of volcanic ash and gas into the stratosphere. The eruption generated a tsunami. This event is rated as a very large volcanic eruption which caused severe destruction in Tonga and several countries around the world. In Tonga the tsunami waves were around 20 m high whereas in Tofua island they were around

TABLE 1.3
List of submarine volcanic eruption-generated tsunamis

Sl. No	Year	Volcano name and location	Maximum Run-up	Observations	Reference
1.	27.08.1883	Krakatoa, Indonesia	41 m	36000 death tolls and hit the Madras port with 2m run-up height	Bapat (1982)
2.	18.03.1888	Ritter Island, Papua New Guinea	20 m	1500–3000 deaths	Global Volcanism
3.	22.12.2018	Anak Krakatoa, Indonesia	13 m	437 people were reported death	Walter et al. (2019)
4.	15.01.2022	Hunga Tunga	20 m	Displacing more than 1500 people and causing four deaths	USGS

45 m high. This event caused 1.2-m high waves in the Tongan capital Nuku'alofa. There was extensive ash produced that blocked the sunlight. The volcanic explosion was succeeded by a tsunami roughly 15 minutes later with the first wave being the most significant (Terry et al., 2022).

1.3.4 Meteorite Impact

Fortunately for humanity, the occurrence of large-scale meteorite or asteroid impacts on Earth is exceedingly rare. While there is no documented evidence of a tsunami being generated by an asteroid impact, the potential consequences of such an event would be catastrophic. The majority of meteorites burn up upon entering Earth's atmosphere, but historical records and the presence of large craters on Earth's surface suggest that significant meteorite impacts have occurred in the distant past. It is conceivable that asteroids may have impacted Earth in prehistoric times, with the last notable event believed to have happened around 65 million years ago during the Cretaceous period. Given the existence of evidence indicating meteorite and asteroid falls on Earth, it is reasonable to infer that such events have also occurred in the oceans and seas, especially considering that four-fifths of our planet is covered by water.

1.4 HISTORICAL TSUNAMIS IN THE INDIAN OCEAN

1.4.1 Andaman-Sumatra Subduction Zone

The Indian Ocean has two sources capable of generating tsunamis: (1) the Andaman-Sumatra subduction zone (ASSZ) situated in the east and (2) the Makran subduction zone located to the west of India (Figure 1.4). The Indian Ocean has a mid-ocean ridge where the Indo-Australian plate and African plate are diverging from Antarctic plate. Most of the earthquakes occurring along these ridges are primarily of the strike-slip type, which do not have the potential to generate tsunamis. The Eastern Indian plate is sub-ducting beneath the Burmese plate along the Andaman Sumatra trench. Significant seismicity in this subduction zone is the main contribution to tsunami generation as these earthquakes are typically associated with the thrust mechanism. The Andaman-Sumatra subduction zone is recognized as one of the most active tectonic plate boundaries globally, exhibiting oblique convergence of 50 mm per year (Stein and Okal, 2005) between the South Asian and India-Australian plates. This convergence extends approximately 5500 km from Myanmar, passing through Sumatra and Java, and reaching towards Australia. The tectonic plates converge at a depth of 5 km beneath the ocean surface at the Sumatra trench. This trench runs parallel to the western coastline of Sumatra, approximately 200 km offshore. At this trench the Indian-Australian plates are subducted and overridden by the Southeast Asia plate, while the Sunda arc continues further north towards the Andaman-Nicobar Islands.

Looking into the list compiled by several researchers (Murthy and Bapat, 1991; Dominey-Howes et al., 2007) of tsunamis in the ASSZ from 326 BCE to 2005 CE, there are more than 70 tsunamis from the Sunda-Arc and a few tsunamis in the Andaman-Nicobar Islands (Table 1.4). The zone has experienced three great earthquakes in a short span of three years: (1) December 26, 2004 Sumatra tsunami of Mw 9.3; (2) Nias earthquake of Mw 8.6 on March 28, 2005, and (3) Bengkulu earthquake of Mw 8.4 on September 12, 2007. After the 2004 Sumatra earthquake, the source process involved in generating these three significant earthquakes has been analyzed by scientists to determine the magnitude estimation, the rupture process and its associated tsunami hazard (Sinadinovski, 2006; Bormann et al., 2007; Singh et al., 2007 and Chlieh et al., 2007).

1.4.1.1 Arakan Tsunami, 1762

The 2004 Indian Ocean Tsunami (IOT) was devastating; however, historical records also indicate that the 1762 CE earthquake in the Arakan Subduction Zone (ASZ), located north of the 2004 epicentre, generated a significant tsunami. This event is one of the earliest documented large

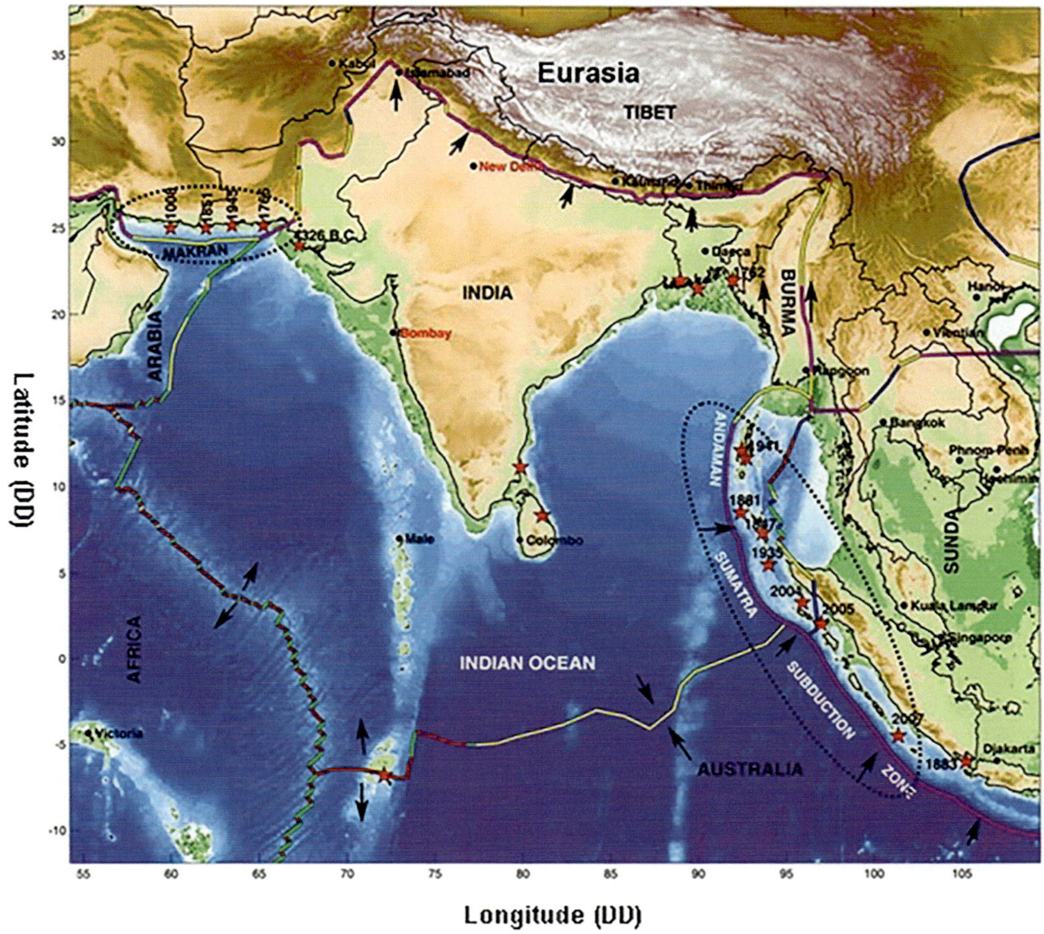

FIGURE 1.4 Map showing the two tsunamigenic sources in the Indian Ocean.

earthquake-induced tsunamis in the northern Bay of Bengal, yet it remains poorly understood due to limited historical evidence. Cummins (2007) estimated the magnitude of the 1762 event at Mw 8.8 as a worst-case scenario and suggested that a similar tsunamigenic earthquake may not occur for more than 200 years. Numerical simulations based on fault parameters and deformation models suggest that although the regional tectonics generally preclude large tsunamis, exceptional events like the 1762 earthquake could still pose a significant hazard. The tsunamigenic earthquake from the historical records has shown that it also generated a significant tsunami (Alam et al., 2012; Aruna Kumari, et al., 2025).

1.4.1.2 Car Nicobar Island Earthquake, December 31, 1881

An earthquake of magnitude Mw 7.9 occurred at Car Nicobar Island on December 31, 1881 at 7.49 am local time with the epicenter at 92.43°E, 8.52°N. A tsunami was generated by this earthquake in the Bay of Bengal. Oritz and Bilham (2003) estimated the rupture characteristics of this earthquake. They gave the fault parameters of this event as: strike angle 360°, Dip 20°, slip magnitude 3.3 m, and focal depth 15 km, and the main rupture occurred between 8.5°E and 10°N with a total area of 150 km × 60 km dipping 20°E with a mean slip of 2.7m. The rupture consisted of two segments; the northern 40-km long segment is separated from the 150-km long segment by a 100-km region corresponding to the westward projection of the west Andaman spreading center. At Port Blair on the

TABLE 1.4
List of significant tsunamis in the Andaman-Sumatra subduction zone

S.No	Date	Location	Long. (DD)	Lat. (DD)	Mag M_w	Observations	Ref.
1	About 500 CE	Poompuhar Tamilnadu (probably due to Krakatau eruption)	79.52	11.12		The town of Poomphar, historically recognized as Kaveipattinam, was a prosperous ancient settlement that was submerged by a tsunami probably caused by the Krakatau eruption	Rastogi and Jaiswal (2006)
3	900 CE	Nagapattinam, Tamilnadu (may be from Sunda-Andaman arc)	79.53	10.46		The Buddhist monastery and several temples were washed away by the tsunami waves, which resulted in loss of hundreds of lives as detailed in Krishnamurty's book "Ponniyin Selvan- The Pinacle of Sacrifice"	Rastogi and Jaiswal (2006)
4	12.04.1762	Bay Of Bengal- Arakan Tsunami	92	22	Appro M_w 8.8	The Arakan tsunami caused extensive devastation and led to 2 m water level rise in the Hoogly River in Kolkata. The sudden rise in water at Dhaka resulted in the loss of hundreds of lives	Mathur (1998); Cummins (2007)
5	11.11.1842	N. Bay Of Bengal	90	21.5		An earthquake along the northern Bay of Bengal triggered a tsunami resulting in the water level rise in the tributaries of the Ganges delta. Boats were violently tossed as if caught in a windstorm	Oldham (1884)
6	31.10.1847	Little Nicobar Island	93.6	7.3	M_w 7.5–7.9	Small island of Kondul located near Little Nicobar was inundated by an earthquake whose Mw, magnitude exceeding 7.5	Berninghausen (1966) Bilham et al. 2005)
7	19.08.1868	Andaman Islands	92.73	11.67			NGDC/NOAA
9	31.12.1881	W of Car Nicobar	92.43	8.52	M_w 7.9	Although the tsunami run-ups and waves heights were not significant their impact was noted in Andaman & Nicobar Islands and along east coast of India	Berninghausen (1966), Ortiz and Bilham (2003)
10	Jan 1882	Sri Lanka (May be from Indonesia)	81.4E	8.34			Berninghausen (1966)
11	1883.08.27	Krakatau (volcanic eruption)	105.25	-6.06		The volcanic eruption of Krakatau in Indonesia resulted in a 35-m-high tsunami claiming 36,000 lives in western Java and southern Sumatra. Tsunami waves traveled transoceanic and effecting several coastal areas	Berninghausen (1966)

12	1884	W. Bay of Bengal				In 1884 a tsunami was observed at Dublet located at the mouth of Hoogly River near Kolkata due to earthquake in the western part of the Bay of Bengal went as far as Port Blair	Murty et al.(1999)
13	31.05.1935	Andaman-Nicobar			M_w 7.5	Tsunami in SW Sumatra	NGDC/NOAA
14	25.11.1935	Andaman-Nicobar	94	5.5	6.5		NGDC/NOAA
15	26.06.1941	Andaman-Nicobar	92.5	12.1	7.7	This tsunami was reported to have wave heights ranging from .75 m to 1.25 m along the east coast of India and approximately 5000 people lost their lives	Bilham et al. 2005
16	26.12.2004	Off West coast of Sumatra and Andaman-Nicobar	95.947	3.30	9.3	The 2004 Sumatra-Andaman earthquake of magnitude 9.3 generated a very large tsunami and killed about 300.000 people in 13 countries situated all around the Indian Ocean	NGDC/NOAA
17	28.03.2005	Nias Region	97.01	2.07	8.7	The tsunami of 2005 caused localized damage with a 3 m wave height damaging Airport on Simeulue. Along west coast of Nias a 2 m wave height was observed	NOAA/NESDIS
18	12.09.2007	Bengkulu Region	101.38	-4.51	8.4	The Benkulu earthquake produced a local tsunami in the vicinity of the epicenter	NGDC/NOAA

Sources: Modified from Rastogi and Jaiswal (2006); Rao (2007); Krishna (2013).

south Andaman Island, a tsunami wave was recorded that was around 1 m height. Literature reveals that first the wave arrived at Nagapattinam at 10.15 am and the observed wave height was about 1.2m. From Nagapattinam the wave propagated north hitting Chennai at 10.18 am, Visakhapatnam in Andhra Pradesh at 11.15 am and then it reached False Point on the Mahanandi delta in Orissa at 11.17 am. From here the waves arrived at Pamban in the Gulf of Mannar at 11.32 am and then progressed towards West Bengal. It was recorded by a tide gauge at Dublat (West Bengal) at 1.00 pm and also at Diamond Harbor (West Bengal) at 3.10 pm. These waves were also observed along the east coast of Sri Lanka. The recurrence time for tsunamigenic earthquakes of this type is estimated to be 114–200 years on the basis of inferred GPS convergence rates (Bilham et al., 2005).

1.4.1.3 Sumatra Earthquake, December 26, 2004

The December 26, 2004 tsunami occurred at 6.29 pm (IST) with an epicenter at 3.307°N 94.5°E with Mw 9.3, and caused large-scale devastation and damage along the coastal regions of India. This earthquake that generated a tsunami has changed the scientific understanding of the hazard globally. The devastating mega thrust earthquake occurred in a tectonically active region where the Indian plate is subducting beneath the Burmese plate in the Sunda trench. This earthquake was felt in Indonesia and globally in countries like Sri Lanka and Africa. This tsunami affected 12 countries from Indonesia to Somalia and killed more than two lakh people and it destroyed billions of dollars of property. The worst affected country was Indonesia as the epicenter of this tsunami was close to this region. To know the reason why this tsunami was strong enough to create havoc, one has to look into the seismological aspects of this earthquake and its rupture characteristics. Numerous studies have been carried out to determine the rupture parameters. The focal mechanism solutions of this event were computed as strike 329°, dip 8°, slip angle/rake 110°, dislocation 20 m and seismic moment 1.0×10^{27} Nm (Lay et al., 2005; Ammon et al., 2005; Choy and Boatwright, 2007), and occurred at a depth of 28.6 km (CMT solution).

The fracture extended unidirectional from Sumatra, moving northward along the plate boundary that runs parallel to the Andaman and Nicobar Islands at the speed of approximately 2.4 km/s for the initial 600km, after which it decreased to 1.5 km/s. The overall rupture zone measures 1400 km, with a maximum displacement of 20 m, most of which occurred within the first 500 km from the epicenter. Notably, a significant study utilized far-field GPS observations, estimating an average reverse slip of 11 m for the southern section of the rupture zone, while the northern section exhibited approximately 10 m (Catherine et al., 2005). The convergence direction of the Indian Plate in relation to the Sunda plate is oriented obliquely to the alignment of the interpolate thrust. Fitch (1972) noted in his research that the Sumatra subduction zone is marked by decoupled faulting.

In larger oceans, such as the Pacific, reflections of the direct tsunami waves are not significant. However, in the Indian Ocean reflections had a notable significant impact, with Kerala experiencing the effect from Lakshadweep Islands. The prolonged elevated water levels in the Andaman and Nicobar Islands may also be attributed to the reflections of the direct tsunami waves and the entrapment of wave energy. The earthquake location was predominantly landlocked on three sides, which meant that instead of the tsunami energy dissipating towards higher northern latitudes, the Bay of Bengal bore the brunt of it. The gradients of ocean depths which led to the convergence and divergence of tsunami wave energy were responsible for the catastrophic waves observed in Sri Lanka. The tsunami traveled and reached the northern region of Sumatra within half an hour following the earthquake, and a few hours later, it impacted Thailand, Sri Lanka, India, and the Maldives. Approximately 10 hours later, the tsunami arrived at the east coast of India.

The tsunami waves impacted various locations along the Indian coast at different times (Nair et al., 2005). They arrived at Andaman and Nicobar Islands approximately around 7.25 am, one hour after the earthquake. Jain et al. (2005) and Murty et al. (2006) examined the tsunami and noted that the tsunami waves reached Andaman and Nicobar islands within 40–50 minutes. The extensive lateral spreading, along with other liquefaction-related phenomenon caused severe damage to pavements, drainage systems, and building foundations. Field studies suggest that low-lying areas and natural features such as sand dunes and mangrove forest played a vital role in dissipating wave energy (Malik et al., 2006; Mascarenhas and Jayakumar, 2008; Ghadamode et al., 2022). The findings of Ghadamode et al. (2024) align with the findings by Cho et al. (2008) and Prerna et al. (2015) for the arrival times and run-up at several locations. The waves reached Chennai at 8.45 am, Velankani coast at 9.05 am, and Kanyakumari around 11.45 am.

The significant loss in Nagapattinam (Tamil Nadu) may be attributed to the resonance of tsunami waves with the natural frequency of the coast line. Subramanyam et al. (2005) demonstrated that the tsunami traveled rapidly across the Bay of Bengal, approaching the continental slope, moving along the shoreline and surging through the bathymetry window between Nagipattnami and Cudadalore. They highlighted the critical role of bathymetry in tsunami amplification along the continental slopes. The wave heights recorded at various locations include approximately 10–15 m along the northwest coast of Sumatra, around 5–10 m on Sri Lanka's east coast, and greater than 5 in the Andaman Islands. Chadha et al. (2005) assessed the tsunami run-up heights by examining water marks on buildings and ocean debris carried inland along the east coast of India and a variation from 2.5 m to 5.2 m with the highest measurement recorded at Nagipattinam. This event from an Indian prospective has been compiled in books by Ramsamy and Kuman (2005) and Murty et al. (2006).

Mahadevan et al. (2005) simulated the waves utilizing the Delft 3D-Flow model which has been developed by WL/Delft Hydraulics in the Netherlands. Elango and Sivkumar (2005) investigated the impact of the tsunami on the groundwater flow in the coastal zones of Tamil Nadu. For anisotropic conditions under non-equilibrium and by solving the groundwater flow through porous media they demonstrated that as saline water recharged the coastal aquifer the salinity of the ground water increased approximately eight times after the tsunami, which is anticipated to decrease as the ground water flows towards the sea. Researchers have quantified and evaluated this natural hazard (Borrero, 2005; Titov, 2006; Geist et al., 2006; Subarya et al., 2006; Stein and Okal, 2005; Srivastava and Dimri, 2007). By using the Geoclaw program, which uses an adaptive finite volume algorithm, Dhanya and Raghukanth (2020) calculated the run-up map for the Indian coast.

1.4.1.4 Nias Earthquake, March 28, 2005

On March 28, 2005 an earthquake occurred in the Nias region at 16:09:36.5 (UTC) with an epicenter at 2.085°N, 97.108°E with a magnitude of Mw 8.6. This earthquake occurred three months after the December 26, 2004 tsunami and also generated a minor tsunami. The epicenter was located near the northern coast of Sumatra, approximately 125 miles west northwest of Sibolga, to the southeast of the 2004 mega tsunamigenic earthquake's epicenter and along the Sunda fault. A tsunami struck both Nias and Simeulle Islands and was a localized event causing destruction in both places. The event was probably triggered by dynamic stress loading caused by the December 26, 2004 (Mw 9.3) earthquake (Knoca et al., 2008). This interaction resulted in convergence at the Sunda Trench, involving oblique movement, which is part thrust-faulting along the plate boundary. The seismically active source region has accommodated some great earthquakes such as 1797, 1833, 1861 of magnitudes greater than 8 and earthquakes with magnitude greater than 7 struck offshore islands in 1881, 1935, 2000, and 2002. For this earthquake the rupture characteristics were estimated by Gahalaut and Catherine (2006) as strike 333°, dip 8°, rake 118°, and dislocation 4–5 m.

1.4.1.5 Bengkulu Earthquake, September 12, 2007

On September 12, 2007 the Sunda mega thrust experienced another earthquake at 16:40:26 (IST) with an epicenter at 4.517°S, 101.382°E and a magnitude of Mw 8.4. This event occurred off the west coast of Sumatra about 130 km SW of Bengkulu and the focal depth of this earthquake was 34 km. The earthquake generated a moderate tsunami because of its size and was followed by minor aftershocks. Twelve hours after this an earthquake of magnitude Mw 7.9 struck near the Sumatra coast, 185 km SSE of Padang. Both earthquakes caused small-scale damage and loss to human life in regions close to the epicenters. The aftershocks continued for two days and on September 13, another earthquake of magnitude Mw 7.1 occurred in the same source region. This region is also seismically active as it experienced two great earthquakes in 1797 and 1833 which generated significant tsunamis. The source parameters of this event have been computed by many researchers using waveform modeling and slip distribution studies (Okal and Synolakis, 2008). This earthquake ruptured 450 km (length) × 200 km (width). The fault plane solution of this event was given as strike 328°, dip 9°, slip angle/rake 114°, dislocation 1–2 m with a seismic moment of 4.2×10^{21} Nm. This event was modeled and assessed by Srivastava et al. (2007) who had said that the Indian coast would be safe from a tsunami and its hazards.

1.4.2 Makran Subduction Zone Earthquakes

The western Indian Ocean is not free from tsunami threats. The Makran subduction zone located on the western side of the Indian Ocean has also a history of tsunamis which have created havoc in the epicentral region and along the west coast of India. Great earthquakes along the Makran subduction zone are very rare but one cannot overlook the seismicity in this region. These events emphasize the need to assess tsunami hazards associated with the tsunamigenic zones in the Indian Ocean (Table 1.5). An earthquake which was recorded, observed, and documented was the November 27, 1945 tsunamigenic earthquake in the Makran subduction zone, which generated a large tsunami in the northern Arabian Sea and caused devastation and destruction in several towns and cities along the west coast of India.

1.4.2.1 Western Makran Earthquake, 1008

A tsunami was triggered and observed in the Indian Ocean due to an earthquake in western Makran in 1008. Documented evidence revealed that this earthquake had generated a tsunami (Murty and Bapat, 1999; Rastogi, 2006). The effects of this tsunami were also documented by Ambraseys and Melville (1982) who gave the location of this earthquake as 27.7°N, 52.7°E. This tsunami sunk many ships and claimed a large number of lives in the Persian Gulf.

1.4.2.2 The Eastern Makran Tsunami, 1765

An earthquake is known to have caused a landslide on the Makran Coast 150 km west of Karachi at some time in the 18th century. The precise identification of 1765 as the year of a great earthquake is obviously inappropriate based on the vague mention of the timing of the earlier event. Ambraseys and Melville (1982) list the year as 1765 with no implied or stated certainty. Byrne et al. (1992) identify this earthquake as a great rupture beneath the easternmost end of the Makran subduction zone. However, a landslide could have also been triggered by a large strike-slip earthquake on the southernmost Ornach–Nal fault system.

1.4.2.3 Makran Tsunami, April 19, 1851

Merewether (1852) reported that on April 19, 1851 an earthquake occurred at 5 pm. This tsunami was observed at Gwadar in Makran and several houses were inundated. He reported three more earthquakes in the same year that were felt in the Gwadar region. This earthquake was later mentioned by Walton (1864) in his letter though the gap is only 13 years.

1.4.2.4 Makran Earthquake, August 25, 1864

On the 25th morning of August 1864 an earthquake occurred along the Makran coast at 00:45 that was felt in Gwadar. There is no documented information about the devastation and loss to the human life and hence was deduced to be a small event ($M<5$). The source of this earthquake is the letter written by Walton (1864).

TABLE 1.5
List of significant tsunamis along the Makran subduction zone

Sl. No.	Year	Longitude (Deg)	Latitude (Deg)	Observations	References
1	326 BCE			Alexander's navy destroyed	Lisitzin (1974)
2	1008 CE	60.0E	25.0N	On the Iranian Coast in North Indian Ocean this tsunami was noticed	Murty et al.(1999)
3	1524	73.2 E	17.0 N	A large earthquake generated the tsunami that caused significant alarm to the Portuguese navy gathered within the region	Bendick and Bilham (1999)
4	1668	68.0 E	24.0 N	With 30,000 houses the Samawani township was sunk into the earth	Oldham (1884)
5	1765	66.5 E	25.1 N	Precipitated a hill into the sea	Walton (1864) Ambraseys and Melville (1982)
6	1819	71.9 E	26.6N	By a terrific rush from the ocean, the town of Sindri (71.9°E, 26.6°N) and neighboring country were flooded and went underwater. The ground sinking observed was about 5 m	Macmurdo (1821)
7	1845	68.37 E	23.6 N	The sea was rolled up the Koree mouth of the Indus spilling over the country as far eastward as to the Sindree lake, northward to the surrounding area of Veyre, and westward to the Goongra river	Nelson (1846)
8	1851	62.0 E	25.0 N		Ambraseys and Melville (1982)
9	1864	62.3 E	25.0 N	Significantly the tsunami carried hundreds of tons of fish to the Makran beaches	Walton (1864)
10	1945	63.0E	24.5 N	Along the western coast of India above 4000 people were killed. Maximum run-up height observed was 17 m	Byrne et al. (1992)

Source: Modified from Rastogi and Jaiswal (2006).

1.4.2.5 Makran Earthquake, November 27, 1945

This earthquake occurred at 21:56 UTC (03:26 IST) on November 27, 1945 and the epicenter of this earthquake was located at 63.0°E, 24.5°N off the Makran coast in the northern Arabian Sea about 100 km south of Karachi and about 87 km SSW of Baluchistan, Pakistan, at a focal depth of 25 km. The size of this earthquake was Mw 8.1 (Beer and Stag, 1946; Byrne et al., 1992). This earthquake was recorded by New Delhi, Kolkatta, and Kodaikanal seismic observatories. Its intensity was high throughout the region and it was strongly felt in Baluchistan and the Las Bela area of Pakistan. There were reports saying that strong tremors were felt which lasted about 30 seconds in the western and southern segments of Karachi (Pakistan). According to the eyewitness reports people were thrown out of their beds, and doors and windows rattled. The underwater cable link between Karachi and Muscat (Oman) was damaged. The lighthouse at Cape Moze, which is 45 km from Karachi, was also damaged. The earthquake was also strongly felt at Manora and moderately felt in Panjaur. This earthquake caused the eruption of a mud volcano a few miles off the Makran coast of Pakistan and this earthquake ruptured the entire eastern segment of the Makran subduction zone. The rupture length was estimated to be about 300–350 km. This earthquake triggered a tsunami and reached a maximum run-up height of 11 m along the Makran coast. There was large-scale devastation,with destruction to property and loss of human lives. The most deaths were reported in Pasni and Ormara and this earthquake was also recorded at Gwadar. Tsunami waves propagated from the epicenter and reached Karachi with a wave height of about 2 m (Pendse, 1948). In Karachi the first tsunami wave was recorded at 5.30 am local time, then at 7.00 am, 7.15 am, and finally at 8.15 am, which was the largest. The direction of this tsunami was from Clifton and Ghizri. The tsunami struck Gujarat and arrived at 8:15 am. It was observed on Salsette Island (Mumbai), Bombay harbor, Versoa (Andheri), Haj Ali (Maha Laxmi), Juhu (Ville Parle), and Danda (Khar).

The convergence between the Eurasian and Arabian plate has led to the formation of the Makran subduction zone which is recognized as one of the accretionary wedges and forearcs on the Earth. This zone displays a notable disparity in the seismicity pattern between its eastern and western regions. It is known to have a shallow subduction angle and the sediment thickness in the area is very high; in some places it is about 7 km. The forearc extending towards the ocean consists of nearly 70 km of newly accreated, semi-consolidated and unconsolidated sediments that exhibit relatively seismic velocities and high pore fluid pressure. The sediments within the accretionary wedge are significantly deformed. The Eastern and Western boundaries of the Makran subduction zone (MSZ) are delineated by the Oranch-Nal fault structure, which is a major strike slip structure, and the Zendan-Minab fault structure (Heidarzadeh et al., 2009), showing the highly complex tectonics in the region. According to Koppa et al. (2000) and Schluter et al. (2002) in contrast to other subduction zones globally, it lacks a trench and exhibits a low dip angle, as evident from recent seismic profiles. The transfer of tectonic stresses between the plates may trigger a tsunamigenic earthquake in the north-western costal area of India in the future. The shallow subduction of the northward moving Omanian oceanic lithosphere beneath the Iranian microplate contributes to the seismo-tectonics of the Makran subduction zone. (Kukowski et al., 2000)

The dynamic process of orogenesis has led to heavily faulted coastal mountain ranges along the Makran coast. The prominent characteristic as noted by Minshull et al. (1992) is the division of MSZ into two segments: the eastern and western segment by the Sistan suture zone. The seismological research has revealed that the eastern Makran experiences mostly thrust earthquakes and is regularly experiencing seismic activity from minor to significant earthquakes (Byrne et al., 1992). The nonactivity of earthquakes in the western section of Makran specifies that either the plate is presently locked or completely aseismic subduction happens at that part. Nevertheless, overall seismicity of the MSZ shows that large and shallow earthquakes are relevant in the MSZ and in future could lead to a tsunamigenic earthquake.

1.5 SUMMARY

Coastal regions are anticipated to see substantial population growth in the coming years and are vulnerable to tsunamis, storms, and floods. This chapter discusses the causes of tsunamis due to earthquakes, land slides, meteorite impact, and submarine volcanoes and describes some of the major tsunamis that have occurred due to these sources. Not all earthquakes generate tsunamis as enormous amounts of energy need to be released during the occurrence of the earthquake to deform the ocean bottom and to uplift the seafloor to significant heights. Thorough maps of probable tsunami inundation and run-up are required for long-term preparation in susceptible coastal communities, and demarcation of evacuation direction is required to mitigate the devastation and destruction that could be caused by tsunamis. To develop the inundation and directivity maps one has to model tsunamis to know the possible wave heights and inundation extents.

REFERENCES

Abe, K. (1979). Size of great earthquakes of 1837–1974 inferred from tsunami data. *Journal of Geophysical Research: Solid Earth*, *84*(B4), 1561–1568. doi: 10.1029/JB084IB04P01561

Abe, K. (1989). Quantification of tsunamigenic earthquakes by the Mt scale. *Tectonophysics*, *166*(1–3), 27–34. doi: 10.1016/0040-1951(89)90202-3

Alam, E., Dominey-Howes, D., Chagué-Goff, C., & Goff, J. (2012). Tsunamis of the northeast Indian Ocean with a particular focus on the Bay of Bengal region—A synthesis and review. *Earth-Science Reviews*, *114*(1–2), 175–193. doi: 10.1016/j.earscirev.2012.05.002

Ambraseys, N. N., & Melville, C. P. (1982). *A history of Persian earthquakes*, Cambridge University Press, London. doi: 10.1016/j.earscirev.2012.07.002

Ammon, C. J., Ji, C., Thio, H. K., Robinson, D., Ni, S., Hjorleifsdottir, V., ... & Wald, D. (2005). Rupture process of the 2004 Sumatra-Andaman earthquake. *Science*, *308*(5725), 1133–1139. doi: 10.1126/science.1112260

Aruna Kumari K., Ghadamode, V., Pandey, A. K., & Srivastava, K. (2025). Tsunami modeling for a possible Tsunamigenic earthquake from the 1762 Arakan source region and its effect along the South Andaman Coast. *Journal of the Geological Society of India*, *101*(6), 866–871. doi: 10.17491/jgsi/2025/174175

Bapat, A. (1982). Assessment of historical seismic data for seismic risk evaluation in India. In *Proceedings of the seventh symposium earthquake engineering* (Vol. 1, pp. 121–124). Springer, Cham.

Baptista, M. A., Miranda, P. M. A., Miranda, J. M., & Victor, L. M. (1998). Constrains on the source of the 1755 Lisbon tsunami inferred from numerical modelling of historical data on the source of the 1755 Lisbon tsunami. *Journal of Geodynamics*, *25*(1–2), 159–174. doi: 10.1016/s0264-3707(97)00020-3

Beer, A., & Stag, J. M. (1946). Seismic sea wave of November 27th (1945). *Nature*, *158*, 63–64.

Bendick, R., & Bilham, R. (1999). Search for buckling of the southwest Indian coast related to Himalayan collision. *Geological Society of America, Special Paper*, 328.

Berninghausen, W. H. (1966). Tsunamis and seismic seiches were reported from regions adjacent to the Indian Ocean. *Bulletin of the Seismological Society of America*, *56*(1), 69–74.

Bilham, R., Engdahl, E. R., Feldl, N., & Satyabala, S. P. (2005). Partial and complete rupture of the Indo-Andaman plate boundary 1847–2004. *Seismological Research Letters*, 76, 299–311. doi: 10.1785/0120020093

Bormann, P., Liu, R., Ren, X., Gutdeutsch, R., Kaiser, D., & Castellaro, S. (2007). Chinese national network magnitudes, their relation to NEIC magnitudes, and recommendations for new IASPEI magnitude standards. *Bulletin of the Seismological Society of America*, *97*(1B), 114–127. doi: 10.1785/0120060078

Borrero, J. C. (2005). Field data and satellite imagery of Tsunami effects in Banda Aceh. *Science, 308*, 1596. doi: 10.1126/science.1110957

Bryant, E. A., & Nott, J. (2001). Geological indicators of large tsunami in Australia. *Natural Hazards*, *24*, 231–249.

Byrne, D. E, Lynn, R. S., & Davis, D. M. (1992). Great trust earthquakes and aseismic slip along the plate boundary of the Makran Subduction zone. *Journal of Geophysical Research*, *97*(B1), 449–478. doi: 10.1029/91/B02165

Catherine, J. K., Gahalaut, V. K., & Sahu, V. K. (2005). Constraints on rupture of Dec 26, 2004 Sumatra earthquake from far field GPS observations. *Earth and Planetary Science Letters, 273*, 673–679. doi: 10.1016/j.epsl.2005.07.012

Chadha. R. K., Latha, G., Harry, Y., Peterson, C., & Katada, T. (2005). The tsunami of the great Sumatra earthquake of *M* 9.0 on 26 December 2004 – Impact on the east coast of India. *Current Science*, *88*(8), 1297–1301.

Chlieh, M., Jean-Philippe, A., Vala, H., Teh-Ru, A. S., Ji, C., Sieh, K., Anthony, S., Helene, H., Linette, P., Yehuda, B., & John, G. (2007). Coseismic slip and after slip of the Great Mw 9.15 Sumatra–Andaman earthquake of 2004. *Bulletin of the Seismological Society of America*, *97*(1A), 152–173. doi: 10.1785/0120050631

Cho, Y. S., Lakshumanan, C., Choi, B. H., & Ha, T. M. (2008). Observations of run-up and inundation levels from the teletsunami in the Andaman and Nicobar Islands: A field report. *Journal of Coastal Research*, *24*(1), 216–223. doi:10.2307/30134248.

Choy, G. L., & Boatwright, J. (2007). The energy radiated by the 26 December 2004 Sumatra–Andaman earthquake estimated from 10-minute P-wave windows. *Bulletin of the Seismological Society of America*, *97*(1A), S18–S24. doi: 10.1785/0120050623

Cummins, P. R. (2007). The potential for giant tsunamigenic earthquake in Northern Bay of Bengal. *Nature, 449*,75–78, doi:10.1038/nature06088

Dawson, A. G., & Stewart, I. S. (2007). Tsunami deposits in the geological record. *Sedimentary Geology, 200*(3–4), 166–183. doi: 10.1016/j.sedgeo.2007.01.002

Dhanya, J., & Raghukanth, S. T. G. (2020). Deterministic tsunami hazard map for India. *Current Science, 119*(10), 1641–1651. doi: 10.18520/cs/v119/i10

Dominey-Howes, D., Cummins, P., & Burbidge, D. (2007). Historic records of teletsunami in the Indian Ocean and insights from numerical modeling. *Nature Hazards, 42*, 1–17. doi: 10.1007/s11069-006-9042-9

Elango, C. L., & Sivakumar, C. (2005). Numerical modeling of effect of Tsunami on ground water flow and Solute transport. In S. M. Ramasamy & C. J Kumanan (Eds.), *Tsunami: The Indian context* (pp. 221–229). Allied Publishers Ltd, Chennai, India.

Fine, I. V., Rabinovich, A. B., Bornhold, B. D., Thomson, R. E., & Kulikov, E. A. (2005). The Grand Banks landslide-generated tsunami of November 18, 1929: Preliminary analysis and numerical modeling. *Marine Geology, 215*(1–2), 45–57. doi: 10.1016/j.margeo.2004.11.007

Fitch, T. J. (1972), Plate convergence, transcurrent faults and internal deformation adjacent to Southeast Asia and the western Pacific. *Journal of Geophysical Research, 77*, 4432–4460. doi: 10.1029/JB077i023p04432

Fujii, Y., Satake, K., Sakai, S. I., Shinohara, M., & Kanazawa, T. (2013). Tsunami source of the 2011 off the Pacific coast of Tohoku Earthquake. *Earth, Planets and Space, 63*(7), 815–820. doi: 10.5047/eps.2011.06.010

Gahalaut, V. K., & Catherine, J. K. (2006). Rupture characteristics of 28 March 2005 Sumatra earthquake from GPS measurements and its implication for tsunami generation. *Earth and Planetary Science Letters, 249*(1–2), 39–46. doi: 10.1016/j.epsl.2006.07.015

Geist, E., Titov, V., & Synolakis, C. (2006). Tsunami: Wave of change. *Scientific American, 294*, 56–63. doi: 10.1038/scientificamerican0106-56

Ghadamode, V., Kondarathi, A. K., Pandey, A. K., & Srivastava, K. (2024). Shoreline and land use–land cover changes along the 2004-tsunami-affected South Andaman coast: Understanding changing hazard susceptibility. *Natural Hazards and Earth System Sciences, 24*(9), 3013–3033. doi 10.5194/nhess-24-3013-2024

Ghadamode, V., Srivastava, K., Singh, R. K., & Pandey, A. K. (2022). Spatial analysis techniques for tsunami vulnerability and inundation mapping of Andaman region using remote sensing, GIS, AHP, and Fuzzy logic methods. *Environmental Earth Sciences, 81*(17), 427. doi: 10.1007/s12665-022-10548-w

González-Vida, J. M., Alvear-Flores, G., & Medina-Cetina, Z. (2019). Early warning system of earthquake-triggered tsunamis using multi-model ensembles: The case of the Mexican Pacific. *Frontiers in Earth Science, 7*, 242. doi: 10.3389/feart.2019.00242

Gupta, H. K., & Gahalaut, V. K. (2013). *Three great Tsunamis: Lisbon (1755), Sumatra-Andaman (2004) and Japan (2011)*. Springer, London.

Gupta, H. K. (2005). Mega-Tsunami of 26th December, 2004: Indian initiative for early warning system and mitigation of oceanogenic hazards. *Episodes, 28*(1), 2–5.

Heidarzadeh, M., Pirooz, M. D., & Zaker, N. H. (2009). Modeling the near-field effects of the worst-case tsunami in the Makran subduction zone. *Ocean Engineering, 36*(5), 368–376.

Higman, B., Shugar, D., Stark, C., Ekström, G., Koppes, M.N., Lynett, P., Dufresne, A., Haeussler, P. J, Geertsema, M., Gulick, S., Mattox, A., Venditti, J.G., Walton, M.A.L., McCall, N., McKittrick, E., MacInnes, B., Bilderback, E.L., Tang, H., Willis, M.J., Richmond, B., Reece, B., Larsen, C., Olson, B., Capra, J., Ayca, A., Bloom, C., Williams, H., Bonno, D., Weiss, R., Keen, A., Skanavis, V., & Loso, M. (2018). Glacier retreat and the largest landslide-triggered marine tsunami since 1958. *Science Advances, 8*, 12993. doi: 10.1038/s41598-018-30475-w.

Ichinose, G., Somerville, P., Thio, H. K., Graves, R., & O'Connell, D. (2007). Rupture process of the 1964 Prince William Sound, Alaska, earthquake from the combined inversion of seismic, tsunami, and geodetic data. *Journal of Geophysical Research: Solid Earth, 112*(B7), 4728. doi: 10.1029/2006JB004728

Jain, S. K., Murty, C. V. R., Rai, D. C., Malik, J. N., Sheth, A., & Jaiswal, A. (2005). Effects of M9 Sumatra earthquake and tsunami of December 26, 2004. *Current Science, 88*(3), 357–359.

Kanamori, H. (1977). Seismic and aseismic slip along subduction zones and their tectonic implications. *Island Arcs, Deep Sea Trenches and Back-Arc Basins, 1*, 163–174. doi: 10.1029/ME001p0163

Konca, A. O., Avouac, J. P., Sladen, A., Meltzner, A. J., Sieh, K., Fang, P., ... & Helmberger, D. V. (2008). Partial rupture of a locked patch of the Sumatra megathrust during the 2007 earthquake sequence. *Nature, 456*(7222), 631–635. doi: 10.1038/nature07572

Koppa, C., Fruehn, J., Flueh, E. R., Reichert, C., Kukowski, N., Bialas, J., & Klaeschen, D., (2000). Structure of the Makran subduction zone from wide-angle and reflection seismic data. *Tectonophysics*, *329*, 171–191. doi: 10.1016/S0040-1951(00)00195-5

Krishna, K. (2013). *Inundation studies for a few nuclear power plants and other major cities along the east and west coast of India due to tsunamigenic earthquakes in the Indian Ocean*. PhD Thesis, Osmania University, Hyderabad, India.

Kukowski, N., Schillhorn, T., Flueh, E. R., & Huhn, K. (2000). Newly identified strike-slip plate boundary in the northeastern Arabian Sea. *Geology*, *28*(4), 355–359. doi: 10.1130/0091-7613(2000)28<355:NISPBI>2.0.CO;2

Kumar, T. S., Nayak, S., Kumar, P., Yadav, R. B. S., Kumar, A., Sunanda, M. V., ... & Shenoi, S. S. C. (2012). Performance of the tsunami forecast system for the Indian Ocean. *Current Science*, *102*, 110–114. Doi: 10.1007/s12219-012-0181-1

Lander, J. F. (1996). *Tsunamis affecting Alaska, 1737-1996. NGDC key to geophysical research documentation no. 31*. NOAA National Geophysical Data Center, Boulder, CO.

Lay, T., Kanamori, H., Ammon, C. J., Nettles, M., Ward, S. N., Aster, R. C., ... & Sipkin, S. (2005). The great Sumatra-Andaman earthquake of 26 December 2004. *Science*, *308*(5725), 1127–1133. doi: 10.1126/science.1112250

Lisitzin, E. (1974). Sea-level changes. Elsevier Oceanogr. Ser. 8. Elsevier Sci. Publ. Co., Amsterdam, Oxford, New York, vi + 286 p.

Macmurdo, C. (1821). Account of the earthquake which occurred in India in June 1819. *Edinburgh Philosophical Journal*, *4*, 106–109.

Mahadevan, R., Chandramohan, P., and Van Holland, G., 2005, Hydrodynamics of tsunamis. In S. M. Ramasamy & C. J Kumanan (Eds.), *Tsunami: The Indian context* (pp. 69–77). Allied Publishers Ltd, Chennai, India.

Malik, J. N., Murty, C. V. R., & Rai, D. C. (2006). Landscape changes in the Andaman and Nicobar Islands (India) after the December 2004 great Sumatra earthquake and Indian Ocean tsunami. *Earthquake Spectra*, *22*(3_Suppl), 43–66. doi: 10.1193/1.2206792

Marghany, M. (2014). Simulation of tsunami impact on sea surface salinity along Banda Aceh coastal waters, Indonesia. In M. Marghany (Ed.), *Advanced geoscience remote sensing* (pp. 229–251). Intech, Rijeka. doi: 10.5772/58570

Mascarenhas, A., & Jayakumar, S. (2008). An environmental perspective of the post-tsunami scenario along the coast of Tamil Nadu, India: Role of sand dunes and forests. *Journal of Environmental Management*, *89*(1), 24–34. doi: 10.1016/j.jenvman.2007.01.053

Mathur, S. M. (1998). *Physical geology of India*. National Book Trust of India, New Delhi, India.

Merewether, W. (1852). A report of the disastrous consequences of the severe earthquake felt in the provinces of Upper Scinde on the 24th January 1852. *Journal of the Bombay Geographical Society, 10*, 284–286.

Minshull, T. A., White, R. S., Barton, P. J., & Collier, J. C. (1992). Deformation at plate boundaries around the Gulf of Oman. *Marine Geology*, *104*, 265–277. doi: 10.1016/0025-3227(92)90101-M

Mori, N., Takahashi, T., Yasuda, T., & Yanagisawa, H. (2011). Survey of 2011 Tohoku earthquake tsunami inundation and run-up. *Geophysical Research Letters*, *38*(7). doi: 10.1029/2011GL049210

Murthy, P. S. N., & Bapat, A. (1991). Tsunami: The great waves. *Oceanographic Literature Review*, *10*(38), 705–711.

Murty, C. V. R., Rai, D. C., Jain, S. K., Kaushik, H. B., Mondal, G., & Dash, S. R. (2006). Performance of structures in the Andaman and Nicobar Islands (India) during the December 2004 great Sumatra earthquake and Indian Ocean tsunami. *Earthquake Spectra*, *22*(3_Suppl.), 321–354. doi: 10.1193/1.2742994

Murty, T. S., Bapat, A. and Prasad V., 1999: Tsunamis on the coastlines of India. *Science of Tsunami Hazards*, *17*(3): 167–172.

Murty, T. S., Rao, A. D., Nirupama, N., & Nistor, I. (2006). Numerical modelling concepts for tsunami warning systems. *Current Science*, *90*(8), 1073–1081.

Nair, M. M., Nagarajan, K., Srinivasan, R., & Kanishkan, B. (2005). Inadian Ocean Tsunami 2004 – An Indain perspective. In S. M. Ramaswamy & C. J. Kumanan (Eds.), *Tsunami: The Indian context* (pp. 99–109). Allied Publishers Ltd, Chennai, India.

Nelson, C. (1846). Notice of an earthquake and a probable subsidence of the land in the district of Cutch, near the mouth of Koree, or the eastern branch of the Indus in June 1845. *Quarterly Journal of the Geological Society of London*, *2*, 103.

Okal, E. A., & Synolakis, C.E. (2008). Far-field tsunami hazard from mega-thrust earthquakes in the Indian Ocean. *Geophysical Journal International*, *172*, 995–1015. doi: 10.1111/j.1365-246X.2007.03674.x

Oldham, R. D. (1884). The Cachar earthquake of 10th January 1869. *Memoirs of the Geological Survey of India*, *19*, 1–104.

Ortiz, M., & Bilham, R. (2003). Source area and rupture parameters of the 31 December 1881 Mw=7.9 Car Nicobar earthquake estimated from tsunami recorded in the Bay of Bengal. *Journal of Geophysical Research: Solid Earth*, *108*(B4), 2215. doi: 10.1029/2002JB001941

Panizzo, A., De Girolamo, P., & Di Risio, M. (2005). Numerical study of landslide-generated tsunamis by a coupled model (fluid dynamics and DEM). *Journal of Hydraulic Research*, *43*(5), 488–502. doi: 10.1080/00207120500342639

Park, J., Song, T. R. A., Tromp, J., Okal, E., Stein, S., Roult, G., ... & Rosat, S. (2005). Earth's free oscillations excited by the 26 December 2004 Sumatra-Andaman earthquake. *Science*, *308*(5725), 1139–1144. doi: 10.1126/science.1112305

Pendse, C. G. (1948). The Makran earthquake of the 28th of November, 1945. *Indian Meterological Department of Science Notes*, *10*, 141–145.

Plafker, G., Kachadoorian, R., Eckel, E. B., & Mayo, L. R. (1969). *Effects of the earthquake of March 27, 1964 on various communities*. US Geological Survey Professional paper 542-G. US Geological Survey, Washington, DC.

Prerna, R., Srinivasa Kumar, T., Mahendra, R. S., & Mohanty, P. C. (2015). Assessment of Tsunami hazard vulnerability along the coastal environs of Andaman Islands. *Natural Hazards, 75*(1), 701–726. doi: 10.1007/s11069-014-1336-8

Ramasamy, S. M., Kumanan, C. J., Sivakumar, R., & Singh, B. (2005). India. In S. M. Ramasamy, C. J. Kumanan, R. Sivakumar, & B. Singh (Eds.), *Geomatics in Tsunami* (pp. 107–134). New India Publishing Agency, New Delhi, India. doi: 10.59317/9789

Rao, N. P. (2007). Characterization of potential tsunamigenic earthquake source zones in the Indian Ocean. In A. Kundu (Ed.), *Tsunami and nonlinear waves* (pp. 285–312). Springer, Berlin, Heidelberg. doi: 10.1007/978-3-540-71256-5_14

Rastogi, B. K. (2006). A catalog of Tsunamis in the Indian Ocean. In T. S. Murty, U Aswathanarayana and N. Nirupama (Eds.), *The Indian Ocean Tsunami* (pp. 3–18). Taylor and Francis Group.

Rastogi, B. K., & Jaiswal, R. K. (2006). A catalog of tsunamis in the Indian Ocean. *Science of Tsunami Hazards*, *25*(3), 128–143.

Schluter, H. U., Prexl, A., Gaedicke, C., Roeser, H., Reichert, C., Meyer, H., & von Daniels, C. (2002). The Makran accretionary wedge: Sediment thicknesses and ages and the origin of mud volcanoes. *Marine Geology*, *185*, 219–232. doi: 10.1016/S0025-3227(02)00192-5

Siegel, F. R.(2020). Structures that protect coastal populations, assets, and GDPs: Sea dikes, breakwaters, seawalls. In *Adaptations of coastal cities to global warming, sea level rise, climate change and endemic hazards* (pp. 11–25). Springer. doi: 10.1007/978-3-030-22669-5_3

Sinadinovski, C. (2006). The event of 26th of December 2004- The biggest earthquake in the world in the last 40 years. *Bulletin of Earthquake Engineering*, *4*, 131–139. doi: 10.1007/s10518-006-9002-5

Singh, S. K., Shapiro, N. M., Nakata, N., Araki, E., Bialas, J., & Suyehiro, K. (2007). The energy and rupture process of the 2004 Sumatra-Andaman earthquake from observations of the sea floor and at the land surface. *Science*, *315*(5822), 1679–1683. doi: 10.1126/science.1112260

Srivastava, K., & Dimri, V. P. (2007). Tsunami propagation of the 2004 Sumatra earthquake and fractal analysis of the aftershock activity. *Indian Journal of Marine Sciences, 36*(2), 128–135.

Srivastava, K., Swaroopa, V., Srinagesh, D., & Dimri, V. P. (2007). Could the 12 Sept, 2007, earthquake of southern Sumatra, Indonesia, have generated a large Tsunami causing damage to the east coast of India? *Current Science*, *93*(9), 1228–1229.

Stein, S., & Okal, E. A. (2005). Speed and size of the Sumatra earthquake. *Nature, 434*, 581. doi: 10.1038/434581a

Subarya, C., Chlieh, M., Prawirodirdjo, L., Avouac, J. P., Bock, Y., Sieh, K., ... & McCaffrey, R. (2006). Plate-boundary deformation associated with the great Sumatra–Andaman earthquake. *Nature*, *440*(7080), 46–51.doi: 10.1038/nature04522

Subramanyam, C. S., Girish, R., & Gahalaut, V. (2005). Continental slope characteristics along the Tsunami affected areas of Eastern offshore of India and Sri Lanka. *Journal of Geological Society of India, 65*, 778–780.

Sun, Y., & Sun, J. (2020). Self-assessment of tsunami evacuation logistics: Importance of time and earthquake experience. *Transportation Research Part D: Transport and Environment*, *87*, 102512. doi: 10.1016/j.trd.2020.102512

Synolakis, C. E., & Okal, E. A. (2002). Overview of the international decade for natural disaster reduction. *Natural Hazards*, *25*(1), 195–207.

Tappin, D. R., Matsumoto, T., & Watts, P. (2008). The Sissano, Papua New Guinea tsunami of July 1998: Offshore evidence on the source mechanism. *Marine Geology*, *175*(1–4), 1–23. doi: 10.1016/s0025-3227(01)00131-1

Terry, J. P., Goff, J., Winspear, N., Bongolan, V. P., & Fisher, S. (2022). Tonga volcanic eruption and tsunami, January 2022: Globally the most significant opportunity to observe an explosive and tsunamigenic submarine eruption since AD 1883 Krakatau. *Geoscience Letters*, *9*(1), 24.

Titov, V. (2006). Sumatra Tsunami: Lessons from modeling. *Surveys in Geophysics,* 27, 679–705. doi: 10.1007/s10712-006-9012-5

Van Dorn, W. G. (1987). Tide gage response to tsunamis. Part II: Other oceans and smaller seas. *Journal of Physical Oceanography*, *17*(9), 1507–1516.

Walton, H.I., 1864, Correspondence, 27 August 1864. *Proceedings of the Geographical Society Bombay, 17*, cxxv.

Walter, T. R., Haghighi, H., Mahmud, S., Felix, M., Coppola, D., Motagh, M., Saul, J., Babeyko, A., Dahm, T., Troll, V. R., Tilmann, F., Heimann, S. (2019). "Complex hazard cascade culminating in the Anak Krakatau sector lapse". *Nature Communications*, *10*(1), 4339. doi: 10.1038/s41467-019-12284-5

Whitmore, P., Benz, H., Bolton, M., Crawford, G., Dengler, L., Fryer, G., ... & Wilson, J. (2008). NOAA/West coast and Alaska tsunami warning center Pacific Ocean response criteria. *Science of Tsunami Hazards*, *27*(2), 1–19.doi: 10.15389/scihazards.27.2.1-19

2 Mathematical Formulation of the Three Stages of Tsunami

Generation, Propagation, and Inundation

2.1 INTRODUCTION

Tsunami modeling investigations require the solutions of analytical and numerical methods for forecasting tsunami source generation, propagation, run-up, and inundation. Various groups have obtained a closed-form solution to the strain field in the source region, which gives the deformation at the sea bottom and thereby the initial wave. This is followed by solving the wave equation to compute the wave propagation. These equations are solved numerically to obtain the wave heights, run-ups, and inundation extents. The tsunamigenic earthquakes in the oceans are modeled to compute the hazard associated and quantify the impact on the coastal cities.

2.2 A BRIEF DESCRIPTION OF AVAILABLE TSUNAMI CODES AND SOFTWARE

MOST (Method of Splitting Tsunami): This model, developed by Titov and Gonzalez (1997) of PMEL (Pacific Marine Environmental Laboratory) and Synolakis of the University of Southern California, is the standard model used at the NCTR (NOAA Center for Tsunami Research). This is a collection of numerical simulation programs designed to model the three phases of a tsunami: the earthquake source deformation, the transoceanic propagation, and the flooding of coastal areas. It uses nested computational grids of coarse resolution in the open sea, which are used to model the seafloor deformation and propagation down into the high-resolution grids to assess the hazard in the area of interest. The earthquake location and magnitude are utilized in the ground deformation model to obtain the vertical displacement of the ocean surface, which subsequently serves as the initial condition for the tsunami wave in the MOST model.

ANUGA (ANU [Australian National University], GA [Geoscience Australia]): This is a hydrodynamic modeling code allowing users to model complex scenarios like dam breaks, tsunami inundation, etc. The finite volume method is used for solving the shallow water wave equations in two dimensions. This model uses triangular mesh to define the model region and the governing equations are solved at each cell to follow the water depth and horizontal momentum components as a function of time. It defines complex regions with different grid resolutions and a constant bottom friction of 0.025. The code is used with Mercator projected data. In this code four nesting grids are used with increasing grid resolutions while moving from one stage into the other. ANUGA is a joint development project between Geoscience Australia (GA) and the Australian National University (ANU).

FUNWAVE (Fully Non-linear Wave Equations): This is a phase-resolving, time-stepping Boussinesq model for ocean surface wave propagation in the near shore. In NearCoM, it serves as

DOI: 10.1201/9781003784500-2

a circulation module with phase resolving wave information to drive the sediment module. This is a simulation model for propagation and inundation developed by Wei and Kirby (1995) and Wei et al. (1995) at the University of Delaware. This model operates under fully nonlinear and dispersive conditions. It incorporates bottom dissipation and wave breaking to prevent the amplification of the wave at the coastline. Using the moving shoreline algorithm one obtains the land inundation incorporating short wave shoaling, breaking, and run-up. This code is implemented over a Cartesian coordinate grid and does not take into account Coriolis effects.

SWAN: This is another nonlinear shallow water code for long waves developed by Mader (1988, 2001). It dispenses with the depth to wavelength term and this approximation holds good for the propagation model. The same was exploited in the propagation model which also ignores the vertical acceleration of fluid. This model uses a finite difference scheme with explicit time. The effect of bottom friction is not considered in the model. At various time steps the deduced quantities like that of instantaneous water height with undisturbed sea level is taken as the reference point. Normally global modeling uses geographic coordinates but SWAN code is used with Mercator projected data.

The implementation of the earthquake source modeling under MIRONE includes both the Mansinha and Smylie (1971) and Okada (1985) formulation for translating the co-seismic displacements at the sea bottom to the initial wave. MIRONE is a MATLAB-based multipurpose tool which uses the GMT tools to efficiently define the computation domain and run the tsunami dispersion simulation using the SWAN methodology through a powerful Graphical Unser Interface (GUI). COMCOT (Cornell Multi-grid Coupled Tsunami Model) is a comprehensive tsunami modeling tool that can simulate the entire lifecycle of a tsunami, including its generation, propagation and run-up in the coastal areas. The adaptable nested grid configuration strikes a balance between precision and efficiency. MIKE21, a general purpose commercial code, is a professional software package for the simulation of flows, waves, sediments, and ecology in rivers, lakes, estuaries, bays, coastal areas, and seas. This software is the ultimate Swiss army knife of coastal modeling and has been used for many marine engineering projects.

The TUNAMI-N2 model was developed in Tohoku University and is a numerical analysis model for the investigation of near-field tsunamis. This code was developed for simulating past tsunamis and other possible scenarios of tsunamigenic earthquakes. Imamura (1996) initially developed this code and it was later modified by Yalciner et al. (2005) and Imamura et al. (2006). It uses the finite difference technique based on the Staggered Leap-Frog scheme. A comprehensive description of TUNAMI N2 model and the details of the application procedure can be found in the IOC manual number 35 (Goto et al., 1997). This code is used to model for the tsunamigenic earthquake source to generate the initial wave, followed by propagation, run-up, and inundation.

2.3 GENERATION MECHANISM OF A TSUNAMI DUE TO A SUBMARINE EARTHQUAKE

For a tsunamigenic source the earthquake faulting should be either thrust type or normal, whereas earthquakes with strike-slip mechanism don't generate tsunamis because of the horizontal movement of the blocks. The way of the movement, depth of deformation, length and breadth of the deforming fault, the angle at which the fault plane is dipping, and the slip angle will decide the mechanism of the earthquake as well as the size of the tsunami.

The upthrusted and downthrusted crustal slabs pass on potential energy into the overlying mass of water with radical changes in the sea surface over the affected area. The force passed into the water accumulation results in the generation of a tsunami. There are particular kinds of earthquakes, called tectonic earthquakes, which are associated with the crustal distortion of the earth. The water over the distorted region is evacuated from the balanced situation at the time when tectonic earthquakes take place underneath the ocean. Since the evacuated mass of water, acting under gravitational power, tries to get back to the original balanced condition, then water waves are produced. The fault characterizations have been depicted by Imamura et al. (2006). Because of an underwater earthquake the deformation of the ocean floor occurs and the analytical expressions for that ocean floor deformation were specified by Mansinha and Smilye (1971) and Okada (1985). The ocean floor distortion generates motionless upliftment of the ocean bottom that shapes the necessary initial conditions required for the modeling of the dissemination of a tsunami and its extent of inundation. In close proximity to the source of underwater earthquakes the floor of the sea is uplifted and subsides, pushing the whole column of water upward and downwards. The potential energy that occurs from pushing water higher than the level of mean sea is then transmitted to the flat dissemination of the wave of tsunami. The parameters of the wave of a tsunami are dependent on the parameters of the earthquake fault as the gravity wave produced by the activity of the ocean floor has to disseminate in the highly incompressible fluid. As a result the parameters of the fault play a very important role in the estimation of deformation at the oceanic floor reflected on top of the oceanic plane as sea bottom upliftment. To uplift the ocean bottom vertically the focal point of the earthquake has to be at a relatively shallow depth. The source parameters needed for quantifying the initial deformation of the sea floor because of the movement of the fault are as follows: epicenter of the earthquake, length of the fault, width of the fault, slip angle, dip angle, strike angle, magnitude of slip, and focal depth (Okal, 2006).

Epicenter: The point on the surface of earth, in a straight line over the focal point (or hypocenter) of an earthquake is defined as the epicenter.

Fault length: In a fault plates move relative to each other. This length is measured along the horizontal distance. The total length of a fault is obtained as

$$L = \frac{M_o^{\frac{1}{3}}}{1.935 \times 10^7} km \tag{2.1}$$

$$\text{where Seismic moment } M_o = 10^{\left(\frac{3}{2} \times (M_w + 10.7)\right)} dyne * cm \tag{2.2}$$

and M_w Moment Magnitude of Earthquake is

$$M_w = \frac{2}{3} \log_{10}(M_o - 10.7) \tag{2.3}$$

Fault width: The total width of a fault or fault zone is influenced by rock type, the intensity of the deformation, and its displacement and is given by

$$W = \frac{M_o^{\frac{1}{3}}}{3.87 \times 10^7} km \tag{2.4}$$

Slip: Slip gives the concept of fault behavior and is the relative displacement of rock masses on either side of the fault. It is the comparative movement of an individual block of a fault to the other block and is obtained as

$$\Delta u = \frac{M_o^{\frac{1}{3}}}{6.68 \times 10^8} m \tag{2.5}$$

Strike angle: Strike direction is the intersection of the plane of the fault with the surface of the Earth. It is measured clockwise from north, ranging from 0° to 360°. This angle gives the orientation and the direction of the fault trace and is perpendicular to the dip.

Dip angle: The angle (inclination) between a horizontal plane and bedding plane. Range 0° to 90°. This angle is an important parameter in understanding the fault geometry and the seismic hazard.

Slip angle (rake): The direction of motion is represented by the slip angle (rake), measured from the direction of strike in the fault plane. It gives the hanging wall block movement with respect to the movement of footwall block. Range 0° to ±180°. When, rake > 0°, that means the block of hanging wall moves upward, which indicates the reverse or thrust movement on a fault; on the other hand, when, rake < 0°, it means the hanging wall moves downward, indicating the normal movement on a fault.

Focal depth: The hypocenter or the focus of an earthquake is the location where the accumulated strain energy in the rock is initially released, indicating the point at which the rock first starts to break. This point is located directly below the epicenter at a distance referred to as hypocenter depth or focal depth.

2.4 PROPAGATION: GOVERNING SHALLOW WATER WAVE EQUATION

The wave equations of shallow water are derived by Synolakis and Yalciner (2006) using tsunami hydrodynamics. To derive the Navier-Stokes equations, from which the shallow water wave equations derive, we use tsunami hydrodynamics. We start with Newton's law of motion applied to a small material particle of dimension $\Delta x \times \Delta y \times \Delta z$. The fluid velocity is expressed as

$$\vec{V} = \vec{\mathrm{u}}_{\mathrm{i}} + \vec{\mathrm{v}}_{\mathrm{j}} + \vec{\mathrm{w}}_{\mathrm{k}} \tag{2.6}$$

where u, v, w, represents velocities in X, Y, Z directions, respectively.

The force on a particle is expressed as

$\vec{F} = \text{mass} \times \text{acceleration}$

$= volume \times density \times accleration$

$= \rho \Delta x \Delta y \Delta z \times \frac{DV}{Dt}$

where ρ denotes the density of the fluid.

Force along the X-direction is

F_x = mass x acceleration in X-direction

$$= \rho \Delta x \Delta y \Delta z \times \frac{Du}{Dt} \tag{2.7}$$

As the fluid particle is moving around, the derivative that describes changes in space and time is given as

$$\frac{Du}{Dt} = \lim_{\Delta t \to 0} \frac{u(x+\Delta x, y+\Delta y, z+\Delta z, t+\Delta t) - u(x, y, z, t)}{\Delta t} \tag{2.8}$$

The above derivative is neither a partial derivative nor an ordinary derivative, it is what we refer to in fluids as the material derivative or the derivative following the particle or the advective derivative.

Before finding the derivative $\frac{Du}{Dt}$, let us first consider a simple case: calculate the derivative $\frac{Df}{Dt}$, where f is a function of x and t:

$$\frac{Df}{Dt} = \lim_{\Delta t \to 0} \frac{f(x+\Delta x, t+\Delta t) - f(x,t)}{\Delta t} \tag{2.9}$$

$$f(x+\Delta x, t+\Delta t) \cong f(x,t) + \frac{\partial f}{\partial x}\Delta x + \frac{\partial f}{\partial t}\Delta t \tag{2.10}$$

Using equation (2.10) in equation (2.9) we obtain

$$\frac{Df}{Dt} = \lim_{\Delta t \to 0} \frac{f(x,t) + \frac{\partial f}{\partial x}\Delta x + \frac{\partial f}{\partial t}\Delta t - f(x,t)}{\Delta t}$$

$$= \frac{\partial f}{\partial x}\left[\lim_{\Delta t \to 0} \frac{\Delta x}{\Delta t}\right] + \frac{\partial f}{\partial t}$$

$$= \frac{\partial f}{\partial x}u + \frac{\partial f}{\partial t} \tag{2.11}$$

Using Taylor's expansion we get

$$u(x+\Delta x, y+\Delta y, z+\Delta z, t+\Delta t) \approx u(x,y,z,t) + \frac{\partial u}{\partial x}\Delta x + \frac{\partial u}{\partial y}\Delta y + \frac{\partial u}{\partial z}\Delta z + \frac{\partial u}{\partial t}\Delta t \tag{2.12}$$

Using (2.12) in equation (2.8) we get

$$\frac{Du}{Dt} = \lim_{\Delta t \to 0} \frac{u(x,y,z,t) + \frac{\partial u}{\partial x}\Delta x + \frac{\partial u}{\partial y}\Delta y + \frac{\partial u}{\partial z}\Delta z + z\frac{\partial u}{\partial t}\Delta t - u(x,y,z,t)}{\Delta t}$$

$$= \lim_{\Delta t \to 0} \frac{\frac{\partial u}{\partial x}\Delta x + \frac{\partial u}{\partial y}\Delta y + \frac{\partial u}{\partial z}\Delta z + \frac{\partial u}{\partial t}\Delta t}{\Delta t}$$

$$\Rightarrow \frac{Du}{Dt} = \frac{\partial u}{\partial x}\lim_{\Delta t \to 0}\frac{\Delta x}{\Delta t} + \frac{\partial u}{\partial y}\lim_{\Delta t \to 0}\frac{\Delta y}{\Delta t} + \frac{\partial u}{\partial z}\lim_{\Delta t \to 0}\frac{\Delta z}{\Delta t} + \frac{\partial u}{\partial t}$$

$$\frac{\mathrm{Du}}{\mathrm{Dt}} = \frac{\partial u}{\partial x}\mathrm{u} + \frac{\partial u}{\partial y}\mathrm{v} + \frac{\partial u}{\partial z}\mathrm{w} + \frac{\partial u}{\partial t}$$

$$\frac{Du}{Dt} = u.\frac{\partial u}{\partial x} + v.\frac{\partial u}{\partial y} + w.\frac{\partial u}{\partial z} + \frac{\partial u}{\partial t} \tag{2.13}$$

Using equation (2.13) in equation (2.7) we obtain the Force acting in X-direction as

$$F_x = \rho\Delta x\Delta y\Delta z\left(u\frac{\partial u}{\partial x} + v\frac{\partial u}{\partial y} + w\frac{\partial u}{\partial z} + \frac{\partial u}{\partial t}\right) \tag{2.14}$$

But we know that for every force there is an opposing force which acts in the opposite direction, and we can get that force from pressure. Let P be the pressure then,

$$\mathrm{P} = \frac{\boldsymbol{F}}{\boldsymbol{A}} \tag{2.15}$$

F is the force acting and A is the area
For the X-direction

$$F_x = Pressure\left(P_x\right) \times area\left(A\right) \text{ along with X direction}$$

$$\therefore F_x = \Delta P_x \times \Delta y \Delta z \tag{2.16}$$

Let ΔP be the change of pressure
We know that on any particle the pressure acting in the forward direction must be greater than the pressure acting in the opposite direction, otherwise it is not possible to accelerate the particle in forward direction. Hence the rate of change of pressure in the X-direction is

$$\frac{\Delta P}{\Delta x} \cong \frac{\partial p}{\partial x}$$

$$\text{i.e., } \Delta P_x = \frac{\partial p}{\partial x} \Delta x \tag{2.17}$$

Using equation (2.17) in equation (2.16) we get as

$$F_x = \frac{\partial p}{\partial x} \Delta x \Delta y \Delta z \tag{2.18}$$

This equation to force is equivalent to equation (2.14) but in the opposite direction. Hence we get

$$\rho \Delta x \Delta y \Delta z \left(u\frac{\partial u}{\partial x} + v\frac{\partial u}{\partial y} + w\frac{\partial u}{\partial z} + \frac{\partial u}{\partial t} \right) = -\frac{\partial p}{\partial x} \Delta x \Delta y \Delta z$$

$$=> \rho \left(u\frac{\partial u}{\partial x} + v\frac{\partial u}{\partial y} + w\frac{\partial u}{\partial z} + \frac{\partial u}{\partial t} \right) = -\frac{\partial p}{\partial x}$$

$$=> \left(u\frac{\partial u}{\partial x} + v\frac{\partial u}{\partial y} + w\frac{\partial u}{\partial z} + \frac{\partial u}{\partial t} \right) = -\frac{1}{\rho}\frac{\partial p}{\partial x} \tag{2.19}$$

In the y direction we have

$$\left(u\frac{\partial v}{\partial x} + v\frac{\partial v}{\partial y} + w\frac{\partial v}{\partial z} + \frac{\partial v}{\partial t} \right) = -\frac{1}{\rho}\frac{\partial p}{\partial y} \tag{2.20}$$

In the Z-direction we have to consider the gravitational force along with the rate of change of pressure.

The pressure difference along the Z-direction is

$$\Delta P_z = -\left(\frac{\partial p}{\partial z}\Delta z + \rho g \Delta z\right)$$

$$= -\Delta z\left(\frac{\partial p}{\partial z} + \rho g\right) \qquad (2.21)$$

∴ Force along the Z-direction is
F_z = Pressure × area of cross section of ΔxΔy

$$= -\left(\frac{\partial p}{\partial z} + \rho g\right)\Delta x \Delta y \Delta z$$

$$\therefore \rho \Delta x \Delta y \Delta z\left(u\frac{\partial w}{\partial x} + v\frac{\partial w}{\partial y} + w\frac{\partial w}{\partial z} + \frac{\partial w}{\partial t}\right) = -\left(\frac{\partial p}{\partial z} + \rho g\right)\Delta x \Delta y \Delta z \qquad (2.22)$$

$$\Rightarrow \rho\left(u\frac{\partial w}{\partial x} + v\frac{\partial w}{\partial y} + w\frac{\partial w}{\partial z} + \frac{\partial w}{\partial t}\right) = -\left(\frac{\partial p}{\partial z} + \rho g\right)$$

$$\Rightarrow \left(u\frac{\partial w}{\partial x} + v\frac{\partial w}{\partial y} + w\frac{\partial w}{\partial z} + \frac{\partial w}{\partial t}\right) = -\frac{1}{\rho}\frac{\partial p}{\partial z} - g \qquad (2.23)$$

Equations (2.19), (2.20), (2.23) are known as Euler equations and they describe the motion of all incompressible, inviscid fluids. This means fluids where viscous force is small. For simplicity, let us assume that the vertical acceleration is very small, so that the equation in the Z-direction becomes zero, i.e.,

$$-\frac{1}{\rho}\frac{\partial p}{\partial z} = g$$

$$\Rightarrow \frac{\partial p}{\partial z} = -\rho g \,(\text{known as hydrostatic approximation}) \qquad (2.24)$$

Let $\eta(x, y, t)$ be the amplitude of the wave, then the pressure at any depth Z be equal to,

$$\int_{-d}^{\eta} \frac{\partial p}{\partial z} dz = -\int_{-d}^{\eta} \rho g dz$$

∴ Pressure at depth Z is equal to

$$P = -\rho g [z]_{-d}^{\eta}$$

$$= -\rho g \left[\eta(x, y, t) + \text{d}\right]$$

Pressure along the Z-direction is expressed as

$$=-\rho g\left[\eta(x,y,t)+Z\right] \tag{2.25}$$

Using equation (2.25) in equation (2.19)

$$u\frac{\partial u}{\partial x}+v\frac{\partial u}{\partial y}+w\frac{\partial u}{\partial z}+\frac{\partial u}{\partial t}=-\frac{1}{\rho}\frac{\partial}{\partial x}\left(-\rho g\left[\eta(x,y,t)+Z\right]\right)$$

$$=-\frac{1}{\rho}(-\rho g)\frac{\partial \eta}{\partial x}$$

$$u\frac{\partial u}{\partial x}+v\frac{\partial u}{\partial y}+w\frac{\partial u}{\partial z}+\frac{\partial u}{\partial t}=g\frac{\partial \eta}{\partial x} \tag{2.26}$$

Similarly,

$$u\frac{\partial v}{\partial x}+v\frac{\partial v}{\partial y}+w\frac{\partial v}{\partial z}+\frac{\partial v}{\partial t}=g\frac{\partial \eta}{\partial y} \tag{2.27}$$

There are coupled and nonlinear equations where u, v, w, η are four unknowns and all functions of x, y, z, t. So it is not possible to find the value of those unknowns only by two equations. To reduce the number of variables, let us assume that u, v do not vary much over depth, which is a very realistic assumption for a long water wave equation. To eliminate z we have to integrate along z taking the limit as Z = –d to Z

$$=\eta(x,y,t)$$

i.e., let us linearize (2.26) and (2.27) and these linear problems from the sea floor, which is at depth d (x, y), to the free surface $\eta(x,y,t)$, which is where the wave propagates. Thus the variable z is eliminated, resulting in depth-averaged velocities.
Let us consider the continuity equation, i.e., conservation of mass equation,

$$\frac{\partial u}{\partial x}+\frac{\partial v}{\partial y}+\frac{\partial w}{\partial z}=0 \tag{2.28}$$

Let us consider there is no velocity along the y-direction so that the continuity equation becomes

$$\frac{\partial u}{\partial x}+\frac{\partial w}{\partial z}=0 \tag{2.29}$$

Integrating the continuity equation over the depth, i.e., with respect to z, taking the limit as $z=-d$ to $Z=\eta$ we get

$$\int_{-d}^{\eta}\frac{\partial u}{\partial x}dz+\int_{-d}^{\eta}\frac{\partial w}{\partial z}dz=0$$

$$\therefore \frac{\partial u}{\partial x}[Z]_{-d}^{\eta} = -w[Z]_{-d}^{\eta}$$

$$\frac{\partial u}{\partial x}(\eta) - \frac{\partial u}{\partial x}(-d) = -w(z=\eta) + w(Z=-d)$$

Because w (z = d) = 0, since there is no flow through the bottom boundary at $Z = -d$

$$\therefore \frac{\partial u}{\partial x}(d) + w(Z=\eta) \approx 0$$

$$\frac{\partial u}{\partial x}(d) = -w(Z=\eta) \tag{2.30}$$

By applying the kinematic free surface boundary condition, which requires that a particle on the free surface stays there at the free surface, i.e.,

$$\frac{D\eta}{Dt} = w(x, y, z=\eta, t)$$

$$=> \frac{\partial \eta}{\partial t} \approx w(Z=\eta)$$

$$\text{i.e.,w} = \frac{\partial \eta}{\partial t}$$

$$\therefore \text{d}\ \frac{\partial u}{\partial x} = -\frac{\partial \eta}{\partial t} => \frac{\partial u}{\partial x} = -\frac{1}{d}\frac{\partial \eta}{\partial t} \tag{2.31}$$

We consider there is no velocity along the Y-direction and the term

$$u\frac{\partial u}{\partial x} \ll \frac{\partial u}{\partial t}$$

$$\therefore \frac{\partial u}{\partial t} = g\frac{\partial \eta}{\partial x} \tag{2.32}$$

Taking the derivative with respect to t for equation (2.31) we get

$$\frac{\partial}{\partial t}\left(\frac{\partial u}{\partial x}\right) = -\frac{1}{d}\frac{\partial}{\partial t}\left(\frac{\partial \eta}{\partial t}\right)$$

$$\frac{\partial^2 u}{\partial t \partial x} = -\frac{1}{d}\frac{\partial^2 \eta}{\partial t^2} \tag{2.33}$$

Taking the derivative with respect to x of (2.32) we get

$$=> \frac{\partial}{\partial x}\left(\frac{\partial u}{\partial t}\right) = g\frac{\partial}{\partial x}\left(\frac{\partial \eta}{\partial x}\right)$$

$$=> \frac{\partial^2 u}{\partial x \partial t} = g\frac{\partial^2 \eta}{\partial x^2} \tag{2.34}$$

From equations (2.33) and (2.34) we get

$$g\frac{\partial^2 \eta}{\partial x^2} = -\frac{1}{d}\frac{\partial^2 \eta}{\partial t^2} \quad \text{assuming} \quad \frac{\partial^2 u}{\partial x \partial t} = \frac{\partial^2 u}{\partial t \partial x}$$

$$=> \frac{\partial^2 \eta}{\partial x^2} = -\frac{1}{gd}\frac{\partial^2 \eta}{\partial t^2}$$

$$=> \frac{\partial^2 \eta}{\partial x^2} = \frac{1}{g(-d)}\frac{\partial^2 \eta}{\partial t^2}$$

$$=> \frac{\partial^2 \eta}{\partial x^2} = \frac{1}{gd}\frac{\partial^2 \eta}{\partial t^2}. \quad \text{Since } d = -d(\text{depth})$$

$$\frac{\partial^2 \text{n}}{\partial \text{x}^2} = \frac{1}{C^2}\frac{\partial^2 \eta}{\partial t^2} \tag{2.35}$$

where $C = \sqrt{gd}$ which is the classic wave equation in one dimension and C is denoted as the wave velocity of the (stretched string) wave. Equation (2.35) is also well-known as the D'Alembert solution and is given by

$$\eta(x, t) = F(x - ct) + G(x + ct) \tag{2.36}$$

where F (x − ct) is any function of x − ct and G (x + ct) is any function of x + ct. By applying the condition the specific forms of F and G can only be obtained.
If we take $C^2 = 1$ then (2.35) =>

$$\frac{\partial^2 \eta}{\partial x^2} = \frac{\partial^2 \eta}{\partial t^2} \tag{2.37}$$

The nonlinear shallow water wave equations (NSE), i.e., the two conservation of momentum equations are solved by TUNAMI N2.

$$u\frac{\partial u}{\partial x} + v\frac{\partial u}{\partial y} + \frac{\partial u}{\partial t} = g\frac{\partial \eta}{\partial x} \tag{2.38}$$

$$u\frac{\partial v}{\partial x} + v\frac{\partial v}{\partial y} + \frac{\partial v}{\partial t} = g\frac{\partial \eta}{\partial y} \tag{2.39}$$

where u, v, η are unknowns which are the functions of x, y, t and the conservation of mass equation

$$\frac{\partial u}{\partial x}+\frac{\partial v}{\partial y}=0 \quad (2.40)$$

where the velocities u, v are depth averaged and pressure is calculated with

$$\mathrm{P}(\mathrm{x,y,t})=-\rho g\left[\eta(x,y,t)+Z\right] \quad (2.41)$$

For tsunami waves one assume that the velocities u, v do not vary over depth.

2.5 BOUNDARY CONDITIONS

Open-sea boundary condition:
At the open sea the boundary condition is a free transmission, i.e., it allows the wave to propagate transoceanic regions without any obstruction.

Amplitude, $\eta = +(\sqrt{h}/g)*u$; for advancing waves

$\eta = -(\sqrt{h}/g)*u$; for receding waves

Boundary condition on land:
On land the boundary condition is a perfect reflection (the initial water level η is equal to the ground height *h).*

$$\eta_{i,j}^{n-1}=-h_{i,j}\,(\text{on the land h take -ve sign})$$

Boundary condition at run-up front:
Only in nonlinear computations is run-up taken into consideration; in linear computations run-up is not taken into consideration.
A cell, whether waterless or underwater, can be judged as follows.

$D = h + \eta > 0$, when the cell is considered as underwater and

≤ 0, when the cell is considered as waterless.

Between the waterless and underwater cells the wave front is located. Discharge across the border linking the two cells is carried out if the land altitude in the waterless cell is less than the height of water in the underwater cell. In other scenarios, discharge can be considered as zero.

2.6 DESCRIPTION OF THE TUNAMI-N2 CODE

TUNAMI-N2 code exercises the information (data) that can be formatted into three columns: the X-coordinate which denotes longitude, the Y-coordinate which denotes latitude, and the Z-coordinate (ground altitude is considered as negative and oceanic depth as positive). By using surfer software the data can be reformed into equally spaced grids. We have considered in our study the spacing of a grid of all four (A, B, C, D) grids in a 1:3 ratio, i.e., A (81 arc seconds) and B (27 arc seconds) grids are used to model the linear effects in deep sea and C (9 arc seconds) and D (3 arc seconds) invariable grids are used to model the non-linear effects of the tsunami wave. As the tsunami wave propagates from deep waters to shallow waters the non-linear effects come into the picture, i.e., amplitude dispersion, energy dissipation, bottom friction, and shallow depths, and the program assumes non-linear theory to estimate the run-ups and impact. To simulate the tsunami the code exercises nesting of grids with precise data of topography and bathymetry, i.e., A grid, B grid, C grid, and D grid. We used coarser resolution topography and bathymetry data to model the initial deformation and propagation in order to save the CPU time. Next, to determine the wave with minimum error, nested grids are used to have the lowest quantity of nodes in a wavelength. Comparatively coarse grids are capable of determining the wave in deep water for the reason that its wavelength is long and fewer node points are needed. When the wave move into shallow waters, the length of the wave reduces (and the amplitude increases), which needs more node points, and thereby needs high-resolution grids. We considered the spacing of grids in such a way that it satisfies the Courant-Friedrich-Lewy conditions for checking the convergence of the numerical code to a certain asymptotic limit.

$$\Delta x/\Delta t \leq \sqrt{(2gh_{max})}$$

Δt and Δx denote the temporal and spatial grid sizes, h_{max} is the upper limit of motionless water depth in the computational domain, and g is the gravitational acceleration. We have considered in this study the spacing of the spatial grid by keeping the time step constant, i.e., 15 seconds with a temporal grid spacing of 1 second.

Apart from the bathymetry and topography data one should have precise focal mechanism solutions and fault parameters to compute the source deformation at the time t=0 seconds which is also known as initial sea floor deformation. This code uses Mansinha and Smyile's (1971) deformation model to estimate the seafloor upliftment near the source. TUNAMI-N2 code allows the tsunami wave to propagate freely in the open sea and for that we have to consider the exterior grid (A) in a very large domain as the tsunami propagates transoceanic regions and is interpolated into the grid (B), grid (C), and grid (D). After giving the required inputs the program is complied and executed to get the directivity map, wave amplitudes at different tide-gauge locations, and heights of run-up at different selected positions in the study region. This program generates three outputs and using MATLAB software the results are plotted.

2.7 TSUNAMI RUN-UP AND INUNDATION

As a tsunami moves from the deep water of the open ocean and travels into the shallower coastal waters, as the water depth decreases the tsunami slows down. Despite the reduction in speed the tsunami's energy flux which depends on wave speed and wave height, remains nearly constant. Subsequently the tsunami's speed reduces and its height grows which is called shoaling. Because of this shoaling effect, a tsunami that is unnoticeable in the deep sea, can increase by several meters in height as it approaches the shore.

Tsunamis may reach a maximum vertical height onshore above sea level, often called a run-up height, of tens of meters. Tsunami inundation refers to the movement of seawater into coastal regions. The distance measured horizontally from the furthest point reached by the seawater to the shoreline is known as the inundation distance.

The amplification of shoaling is influenced by the ratio of the group velocity both at the nucleation-site and the coast-site. Tsunami waves pose a threat primarily as they approach the coast. The height of a tsunami wave increases significantly as it enters shallower waters during shoaling processes. Additionally shoaling waves experience a decrease in wavelength. As the wave comes closer to the shore, its height can increase by a factor of two or more, then it will break.

At the coastal region the tsunami wave height can be estimated by using Green's law. Green's law explains that the arithmetical value of the altitude of tsunami on coast is considered by the fourth root of the depth of sea ratio at the forecast point. The depth of the sea on the coastal area, is supposed to be 1 m for the "h" value, however the actual calculated value would be more suitable, if it is obtainable. That means when a tsunami get nearer to the coastline the latter contribution can be ignored. "h" is supposed to be equal to 1 meter and we can obtain the following equation.

$$H = \sqrt[4]{\frac{h_1}{h}} H_1 \rightarrow \sqrt[4]{h_1} H_1$$

where *H1* is the tsunami wave height at the location, *h1* is the water depth at the location, and *H* is the forecast tsunami height.

2.8 SUMMARY

The three steps involved in tsunami modeling are: generation of the initial wave, wave propagation, and wave run-up and inundation. Tsunami modeling requires information regarding the tsunamigenic earthquake source parameters in order to compute the deformation at the source. This would give the initial wave at the source. The second and third stages require the formulation and solution of the wave equation. Several codes for the computation of tsunami generation propagation and inundation have been developed by different researchers. In this chapter the derivation of the wave equation is described along with a brief description of the TUNAMI N2 code used extensively around the world to compute tsunami arrival times and wave heights.

REFERENCES

Goto, C., Ogawa, Y., Shuto, N., & Imamura, F. (1997). *Numerical method of Tsunami simulation with the leap-grog scheme (IUGG/IOC time project).* IOC manual no. 35. UNESCO, London.

Imamura, F. (1996). *Tsunami numerical simulation with the staggered Leap-Frog Scheme, manuscript for TUNAMI code, School of Civil Engineering.* 1–45. Asian Institute of Technology.

Imamura, F., Yalciner, A. C., & Ozyurt, G. (2006). *Tsunami modeling manual.* UNESCO Tsunami modeling course, UNESCO, London.

Mader, C. L. (1988). *Numerical modelling of water waves.* University of California Press, Berkeley, CA.

Mader, C. L. (2001). *Numerical modelling of water waves.* Los Alamos Series in Basic and Applied Sciences. 1988.

Mansinha, L., & Smylie, D. E. (1971). The displacement fields of inclined faults. *Bulletin of the Seismological Society of America, 61*(5), 1433–1440.

Okada, Y. (1985). Surface deformation due to shear and tensile faults in a half-space. *Bulletin of the Seismological Society of America, 75*, 1135–1154. doi: 10.1785/BSSA0750041135

Okal, E. A. (2006). *UNESCO-IOC International Training course on Tsunami numerical modelling, Course-I Tsunami sources and tsunami Propagation.* Personal notes given in UNESCO-IOC.

Synolakis, C., & Yalciner, A. C. (2006). *Derivation of the basic water wave equations.* Tsunami inundation training session, UNESCO – IOC.

Titov, V. V., & Gonzalez, F. I. (1997). *Implementation and testing of the Method of Splitting Tsunami (MOST) model.* NOAA Technical Memorandum ERL PMEL-112. Contribution No. 1927 from NOAA/Pacific Marine Environmental Laboratory.

Wei, G., & Kirby, J. T. (1995). Time-dependent numerical code for extended Boussinesq equations. *Journal of Waterway, Port, Coastal, and Ocean Engineering, 121*(5), 251–261. doi: 10.1061/(ASCE)0733-950X(1995)121:5(251)

Wei, G., Kirby, J. T., Grilli, S. T., & Subramanya, R. (1995). A fully nonlinear Boussinesq model for free surface waves. Part 1: Highly nonlinear unsteady waves. *Journal of Fluid Mechanics, 294*, 71–92. doi: 10.1017/S0022112095002813

Yalciner, A. C., Karakus, H., Ozer, C., & Ozyurt, G. (2005). *Short course on understanding the generation, propagation near and far field impacts of Tsunamis and planning strategies to prepare for future events.* Kuala Lumpur.

3 Tsunami Propagation over Bathymetric Features

3.1 INTRODUCTION

Ocean ridges' bathymetry affects not only the amplitude of tsunamis but also the direction and speed of tsunamis, making bathymetry an essential parameter in tsunami propagation. Titov et al. (2005) demonstrated that the Southwest Indian Ridge and the Mid-Atlantic Ridge acted as waveguides for the tsunami traveling into the Atlantic Ocean. Conversely, the Southeast Indian Ridge, Pacific, Antarctic Ridge, and East Pacific Rise are waveguides for tsunamis entering the Pacific Ocean. Buchwald (1968) pointed out several wave propagation models along the ridge are possible for a uniform ridge. For a synthetic bathymetric ridge model in conjunction with shallow water and the dispersive Variational Boussinesq model, Groesen et al. (2008) carried out the near-coast tsunami wave guiding model.

To evaluate the impact of bathymetry on tsunami propagation in the Pacific Ocean and the Japan Sea, Kenji Satake(1988) used ray tracing of seismic waves for the Spherical Earth model. Yeh et al. (1995) performed a research study on the Shikotan tsunami, which originated in Northern Japan, demonstrating that the Hawaiian Ridge acted as a waveguide for tsunamis traveling from the west, consequently elevating the tsunami wave heights recorded in the Hawaiian Islands. The Emperor Seamount chain, the Hawaiian Ridge, and the Hawaiian Islands are the main features in the North Pacific responsible for tsunami scattering Mofjeld et al. (2004). Later, Marchuk (2009) studied tsunami propagation along the bottom ridges in the Pacific Ocean.

3.2 TSUNAMIS FROM MAKRAN SUBDUCTION ZONE PASSING THROUGH MURRAY RIDGE

There have been several historical tsunamigenic earthquakes in the MSZ due to the convergence of Eurasian and Arabian tectonic plates in the Northern Arabian Sea (Quittmeyer et al., 1979; Murty et al., 1991; Heidarzadeh et al., 2008). This east-west direction MSZ from the Strait of Hormuz in Pakistan has complex tectonic settings (Byrne et al., 1992) and is the largest forearc region of 70–100 Ma. The morpho-tectonic characteristics, contrasting seismicity patterns, and rupture histories of the MSZ, the Sistan suture zone, divide the MSZ into two segments (Minshull et al., 1992). The eastern boundary is the Ornach-Nal fault system, while the western segment is the Zendan-Minab fault. The eastern segment of the Makran subduction zone is more stressed and has had more tsunamigenic earthquakes than the western segment (Swaroopa et al., 2011). Koppa et al. (2000) and Schluter et al. (2002) have shown that the Makran subduction zone is associated with a low dip angle and does not have a trench. Farhoudi et al. (1977) classified it as the largest sedimentary prism accreted throughout the Cenozoic period, and numerous active mud volcanoes are spread along the Makran subduction zone region.

DOI: 10.1201/9781003784500-3

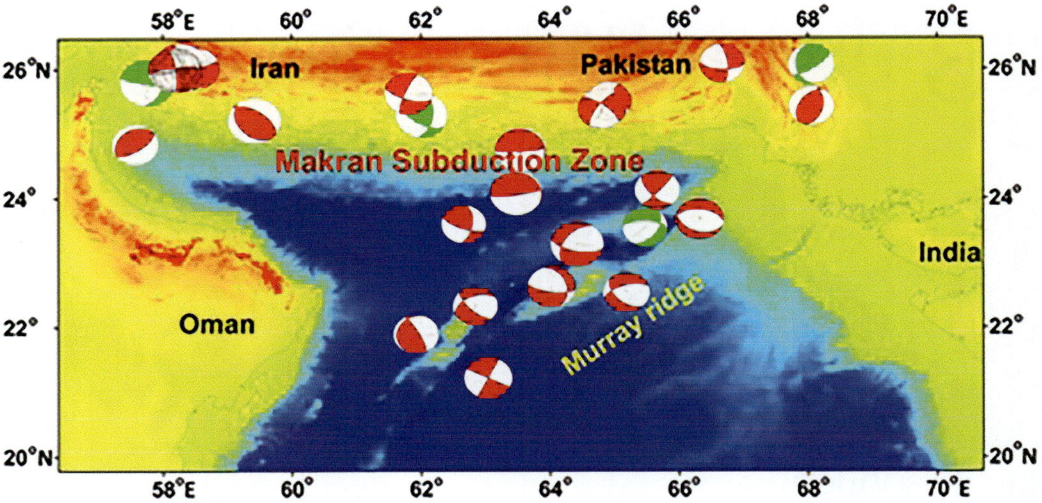

FIGURE 3.1 Seismicity pattern for the Makran subduction zone, which are thrust type, and the Murray Ridge shows strike slip focal mechanism.

The aseismic Murray Ridge in the MSZ is an extension of the Owen Fracture zone, identified by Seymour in 1933 during the John Murray cruise, and is an NE–SW-oriented volcanic ridge. Mita Rajaram et al. (2009) studied India's western continental margin, the corresponding seafloor, and the aseismic Murray Ridge (MR). A 750-km-long ridge separates the Indian and Arabian plates off the coast of Karachi in the northeast, forming a boundary between the two (White, 1984; Edward et al., 2000). There are notable morphological differences between the two segments of the Murray Ridge, the northern and southern MR, along with the Dalrymple Trough. The seismicity of the shallow depths marks the southern MR as a relatively shallow depth of less than 1000 m, while the northern MR reaches a depth of around 2000 m. Gaedicke et al. (2002) have shown that the deepest part is the Dalrymple Trough in the Arabian Sea, having depths up to 4400 m, and its northern part is covered by thick sediments from the Indus fan. Due to its proximity to MR, a tsunami from MSZ will travel through it. To illustrate the impact of the ridge system on tsunami propagation, an analysis is done using the current bathymetry, and the results are compared to a scenario where there is no Murray Ridge bathymetry. The seismicity pattern along the Murray Ridge is shown in Figure 3.1

Removing the bathymetry irregularities along the Makran coast, Neetu et al. (2011) observed the tsunami wave's localization due to the shelf's narrowing. Later, Swapna and Srivastava (2014) studied the tsunami propagation effect with the Murray Ridge and without the ridge system and its effect on the west coast for different tsunamigenic sources. Findings show that a ridge system causes the focus or defocus of the tsunami traveling from MSZ.

3.3 RIDGES ACTING AS A WAVE GUIDE

3.3.1 Case Study: Murray Ridge

A thrust earthquake of magnitude 8.1 in the Makran subduction zone on November 27, 1945 caused significant damage to coastal regions in Baluchistan, Pakistan and the west coast of northern India (Heck, 1947; Quittmeyer et al.,1979; Page et al.,1979). The eastern segment of the MSZ is more stressed and has had several tsunamigenic earthquakes (Swaroopa et al., 2011), while the western segment has not experienced any notable tsunamigenic earthquakes and could be locked. Deterministic and probabilistic methods for tsunamis from the MSZ were studied by Heidarzadeh

TABLE 3.1
Deformation parameters for the November 27, 1945 Makran earthquake

Fault Parameters	November 27, 1945 Makran
Coordinates	63.48^0 E; 25.15^{0}N
Length (km)	120
Width (km)	80
Depth (km)	23
Strike	246
Dip	7
Rake	89
Slip (m)	7
Mw	8.1

and Kijko (2011). In this study for the MR, we considered the source parameters given by Byrne et al. (1992) for the Makran tsunamigenic earthquake (Table 3.1). Pendse (1948) first observed the destruction along the west coast of India and pointed out that at Mumbai the tsunami run-up was around 2 m. Later Ambraseys and Melville (1982) and Murty et al. (1999) carried out hazard analysis along the coast. In this chapter, using the formulation of TUNAMI N2 as discussed in Chapter 2, for the 1945 Makran tsunami, the tsunami arrival times and wave heights are computed.

Using the General Bathymetric Chart of the Oceans (GEBCO), 1 arc minute (1834 m) resolution and topography data from the Shuttle Radar Topography Mission (SRTM), 3 arc sec (93 m) resolution, the tsunami propagation, run-up, and inundation are studied.

The directivity maps, arrival times, and run-up along the west coast are computed for different tsunamigenic earthquakes from MSZ. The tsunami modeling using the bathymetry with natural MRS (Figure.3.2a) is carried out for the fault parameters as given in Table 3.1. The effect of MRS on tsunami propagation with the same set of earthquake parameters is analyzed by taking a constant bathymetry of about 3400 m (Figure.3.2b). Figure 3.3 shows the different profiles L, M, and N considered across the Murray Ridge.

Tsunami waves travel faster in the absence of MR, reach the coast earlier without amplification, and slow down in the presence of the MR. The tsunami wave heights amplify over the MR, as seen in Figure 3.4. The figure shows that the wave heights increase above the Murray Ridge along the M and N profiles. In contrast, we do not observe an increase in the L profile, as the bathymetry is nearly flat. These amplified waves on the ridge delay the tsunami arrival times with greater wave heights.

In contrast, in the absence of the ridge, the tsunami waves travel more rapidly as there is no ridge to delay their propagation, resulting in an earlier arrival at the coast without amplification. Conversely, when the MR is present, the tsunami waves slow down at the MR. The wavelength and wave speed decrease, and the energy per unit area increases, which implies that the wave height increases at the MR. Hence, the amplified wave propagates through the ocean, causing a delayed arrival at the coast and increased wave height. For tsunamis originating from eastern MSZ, the direction of the tsunami propagation shifts after the tsunami traverses the MR, changes towards the southeast direction, and one observes the wave guiding nature and the focusing of tsunami waves towards Achu and Karachi (Figure 3.5). Tsunami directivity is usually perpendicular to the strike of the fault; however, it is observed that the tsunami, after reaching Murray Ridge, changes its directivity of propagation slightly by approximately 5°.

3.4 ISLAND ACTING AS A WAVE BARRIER

Lakshadweep or Laccadive Archipelago is a cluster of Islands around 200 to 440 kilometers off the south-western coast of India. Lakshadweep comes from "Lakshadweepa," meaning "one hundred

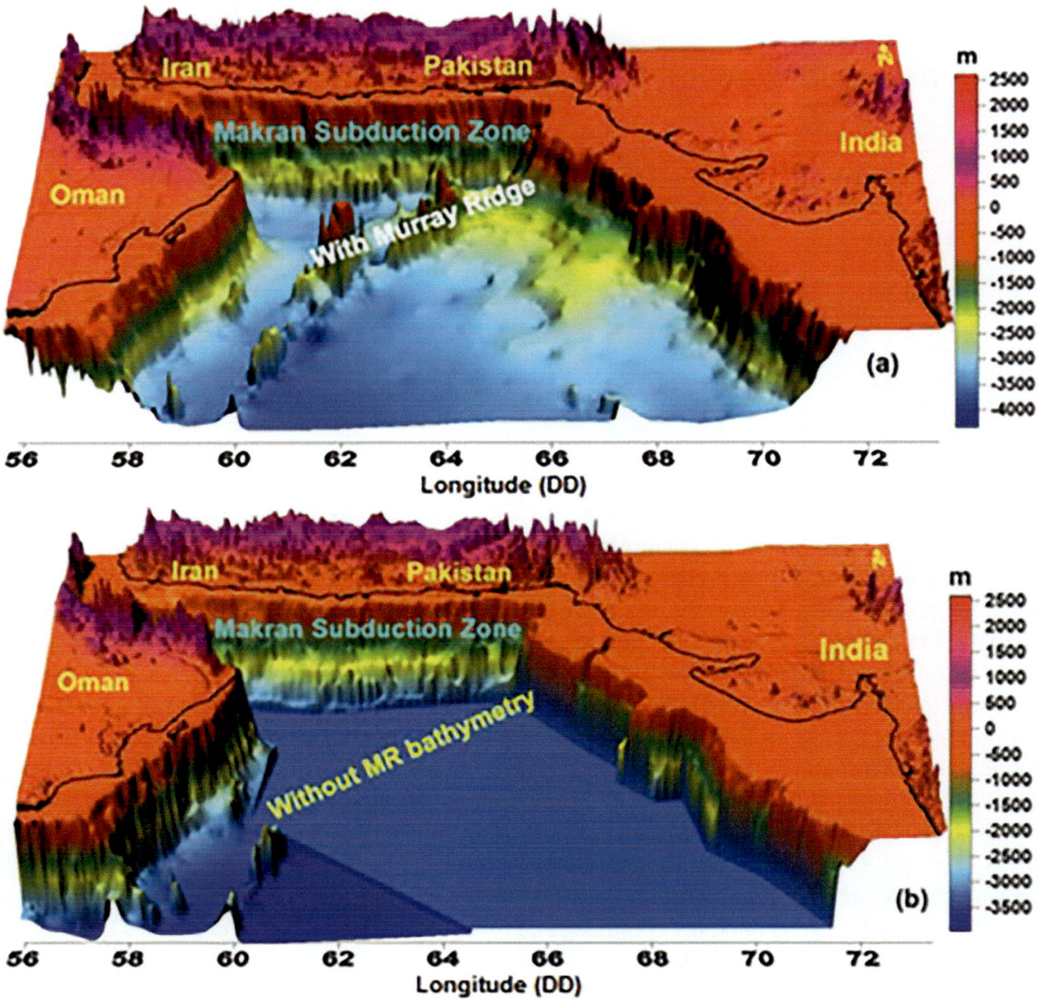

FIGURE 3.2 (a) With Murray Ridge bathymetry. (b) Without Murray Ridge bathymetry.

thousand islands" in Sanskrit. The relief of each island of Lakshadweep is consistently small and ranges from 4 to 5 m. A significant volcanic ridge stretching from the Northern and the Central Indian Ocean is the Chagos-Laccadive Ridge, and the regions exposed above the water are the Lakshadweep islands (Figure 3.6).

The propagation of tsunamis from MSZ into the Northern Indian Ocean and their effect on Lakshadweep's Androth island was carried out by Saha and Srivastava (2016).

For 1945, the source characteristics used in the tsunami simulation are given in Table 3.1. The tsunami source for the 1945 tsunamigenic earthquake is considered.

The deformation at the ocean bottom gives the initial wave, which is seen to be around 3.2 m (Figure 3.7). In this study, the research finding of Shila and Srivastava (2019) is discussed regarding the protection the west coast of India is getting due to the Lakshadweep Islands from the Makran tsunami. Okal (2015) pointed out that during the 1945 Makran tsunami, the main tsunami waves were delayed along the west coast of India.

In the presence of the Lakshadweep islands, the waves amplify with the islands (Figure 3.8), and show an increase in wave heights towards Mangalore. In the absence of the Lakshadweep Islands the directivity of the tsunami waves is towards the open ocean, and an increase in the wave heights

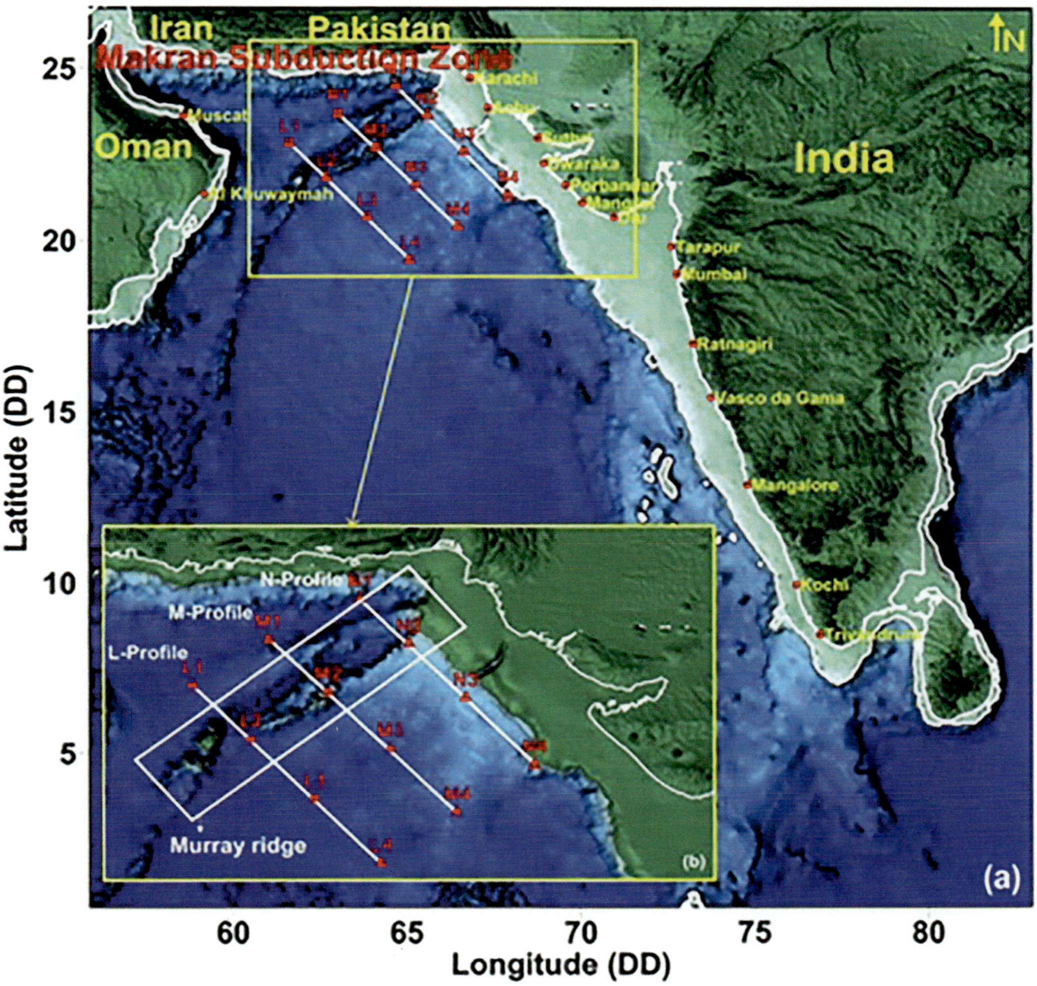

FIGURE 3.3 (a) Tide gauge locations along the west coast of India. (b) Locations taken along L, M, and N profiles across the Murray Ridge.

is observed along the southern part of Mangalore. To illustrate the amplification of the waves, we have examined several locations on and near the Lakshadweep Islands, as depicted in Figure 3.9.

Tsunami propagation in the presence or absence of Lakshadweep does not affect cities like Mumbai, Ratnagiri, and Panjim along west coast of Indian. A change in the wave height is seen from the Mangalore city and down south. The presence of the Lakshadweep Islands caused an increase in wave height at Mangalore and wave height is decreased to the south of Mangalore (Figure 3.9).

We conclude that Lakshadweep acts like a wave barricade or a shield for the southwest coast of India, just as the Ross Islands protected part of Port Blair during the 2004 tsunami (Figure 3.10)

3.5 SHELF WIDTH ALONG THE WEST COAST

With regard to tectonics, the western continental margin of India is a typically passive continental margin and is highly complex hosting a number of deep seated faults, rift systems, basement highs, and NW–SE trending ridges such as Laccadive ridge, Laxmi Ridge, Prathap ridge, etc. The shelf width from north to south along the west coast of India is tabulated in Table 3.2.

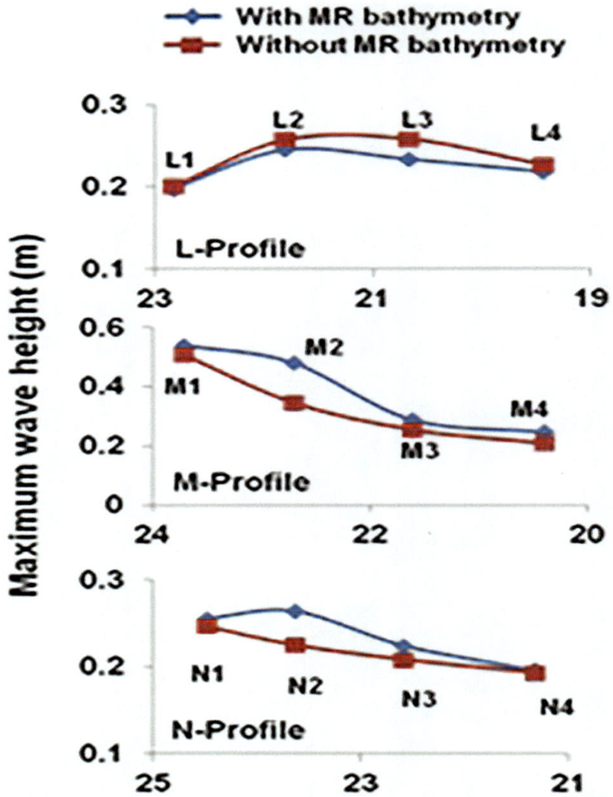

FIGURE 3.4 Estimated maximum wave heights at gauge locations along L, M, and N profiles.

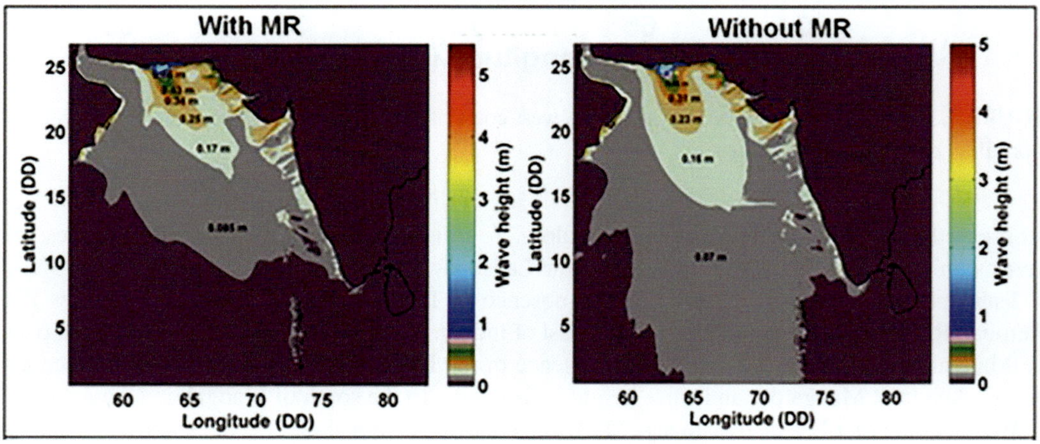

FIGURE 3.5 Directivity map with and without MR.

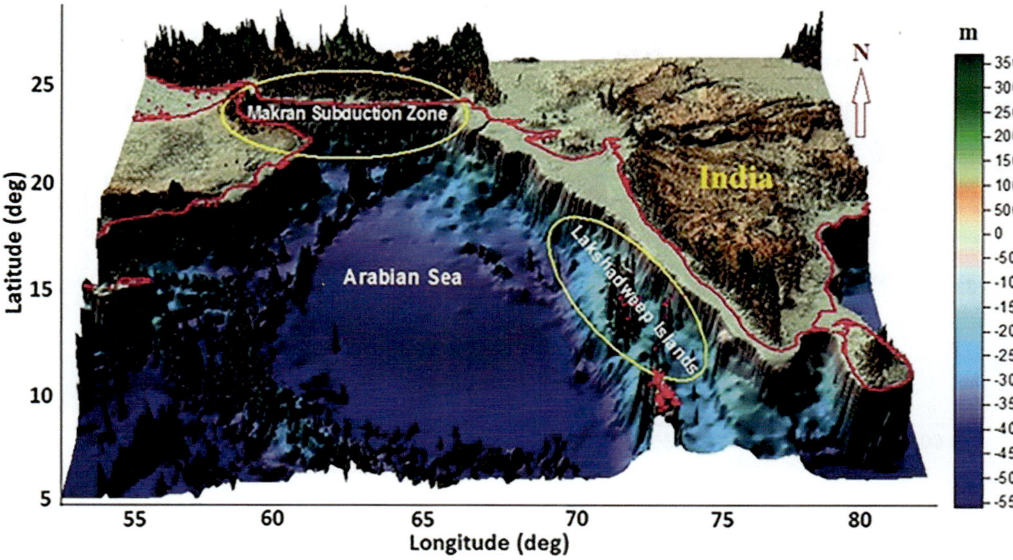

FIGURE 3.6 Makran subduction zone with Lakshadweep Islands. Sources: Saha and Srivastava (2016); Saha and Srivastava (2019).

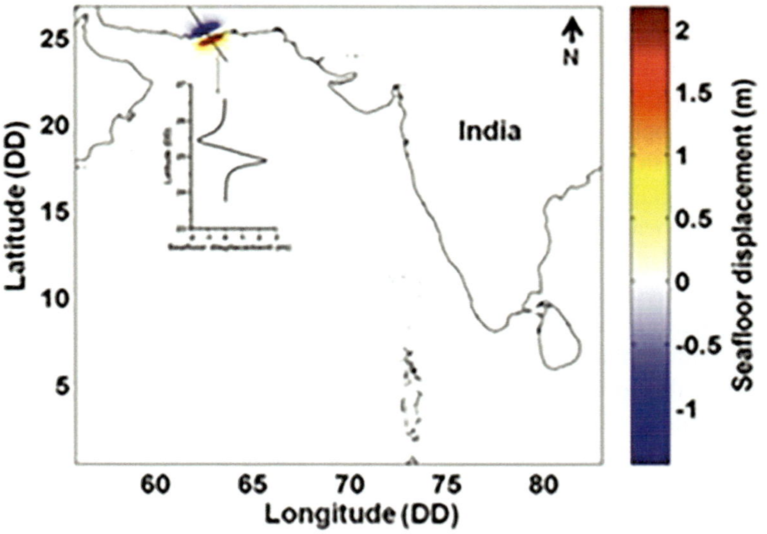

FIGURE 3.7 Deformation at the sea bottom for 1945 Makran tsunami at time t=0.

Tsunami arrival times at various locations on the west coast of India are shown in Figure 3.11. From the figure it is clear that after 106 minutes the tsunami arrived at Dwarka, after 180 minutes Mumbai, after 172 minutes it reached Ratnagiri, after 181 minutes it reached Panaji (Goa), Bhatkal after 213 minutes, Udipi after 209 minutes, Mangalore after 196 minutes, Thalessary after 250 minutes, after 204 minutes it reached Calicut, after 234 minutes Cochin, and after 241 minutes Thiruvanthapuram. The computations show that the tsunami almost simultaneously reaches Mumbai and Panaji (Goa). Surprisingly, the tsunami reached Ratnagir 8 to 9 minutes earlier, 244 km south of Mumbai. At Ratnagiri, the shelf width is narrower, around 125 km, whereas at Mumbai it is 250 km. The fluctuating depth of the water column, caused by the changing width of the shelf in the area, is responsible

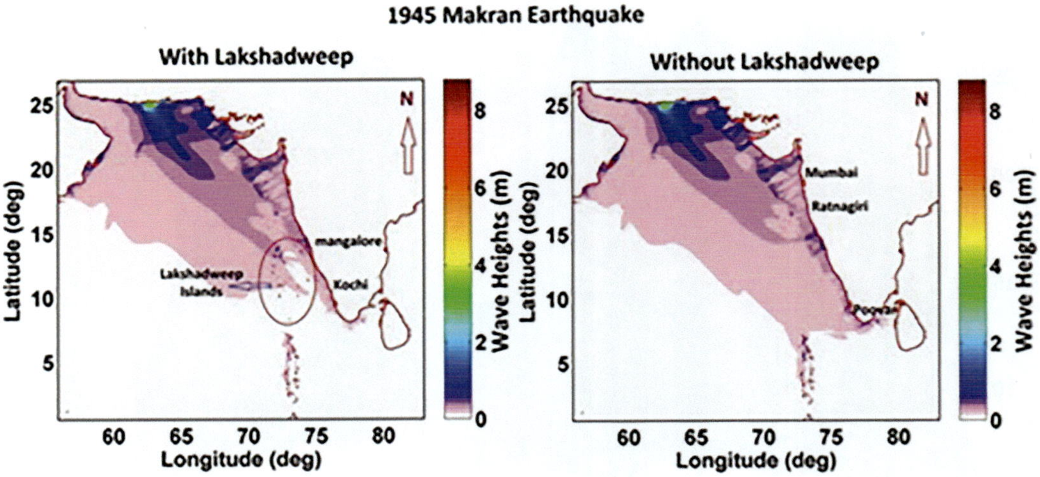

FIGURE 3.8 Directivity map with and without Lakshadweep.

for the difference in the arrival times of the tsunami along the west coast of India. The tsunami's speed is independent of the earthquake's magnitude as $v = \sqrt{gd}$, where v is speed, g is acceleration due to gravity, and d is depth to the water column. A similar observation is noted at Thalessary and Cochin because the depth to the water column varies with the shelf's width. At Bhatkal, the shelf width is about 131 km, and at Udipi, it is 117 km. The tsunami wave reached Udipi after 209 minutes; however, it took longer to reach Bhatkal, at almost 213 minutes. Similarly, the shelf at Cochin is less than at Thalassery and the tsunami reached Cochin 16 minutes earlier. The authorities maintaining the coast should not think that the tsunami has skipped Mumbai and reached a further point down south, and that Mumbai is safe, the tsunami will arrive later. The coast of Mumbai, which is located on reclaimed land, could be highly vulnerable to disasters due to tsunamis. The slumping of sediments near the epicenter region after the earthquake in the MSZ can be attributed to the delays in the arrival at Karachi and Mumbai (Ambaryess and Methwer, 1982; Rajendran et al., 2008). Thus there is a need for a detailed survey of the paleotsunami studies and dating of sediments near the rupture zone. For different scenarios we observe a similar result. Figure 3.12 shows the initial deformation at the source the directivity of the tsunami and the estimated run-up heights near Mumbai and Cochin due to the 1945 Makran tsunami. For different scenarios the tsunami arrival times are computed along the west coast of India and are shown in Figure 3.13 (Srivastava et al., 2011; Swaroopa, 2011).

3.6 TSUNAMIS FROM SUMATRA SUBDUCTION ZONE PASSING THROUGH NINETY EAST RIDGE

The mega thrust tsunamigenic earthquake on December 26, 2004 in the Sumatra region was one of the largest earthquakes ever recorded (Stein and Okal, 2005; Park et al., 2005; Lay et al., 2005). It generated a tsunami, causing devastation throughout the coastal regions surrounding the Indian Ocean.

The seismicity of the Andaman-Sumatra subduction zone, characterized by shallow, intermediate, and deep-focus earthquakes, presents a complex hazard landscape. Most earthquakes in this zone are shallow, occurring along the plate interface, with significant seismic activity also observed along strike-slip faults like the Sumatran fault. These earthquakes' distribution and focal mechanisms provide critical insights into the region's seismic behavior and potential hazards (Figure 3.14a). The Sumatra region is a well-known subduction zone that has already produced the most devastating

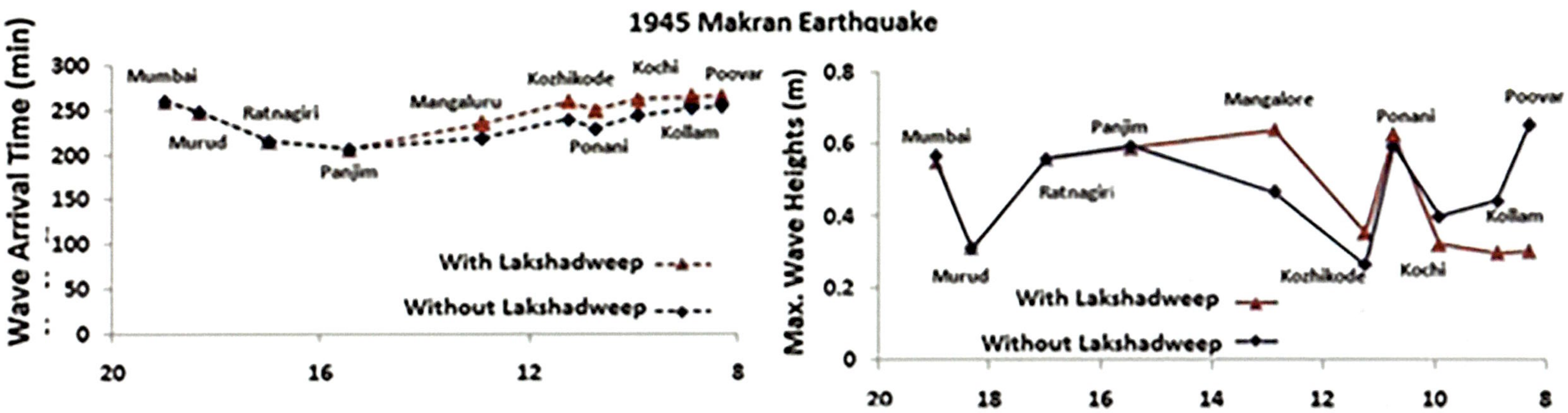

FIGURE 3.9 Estimated wave arrival time and maximum wave amplitude at the selected locations of tide gauge along the west coast of India for 1945 Makran earthquake.

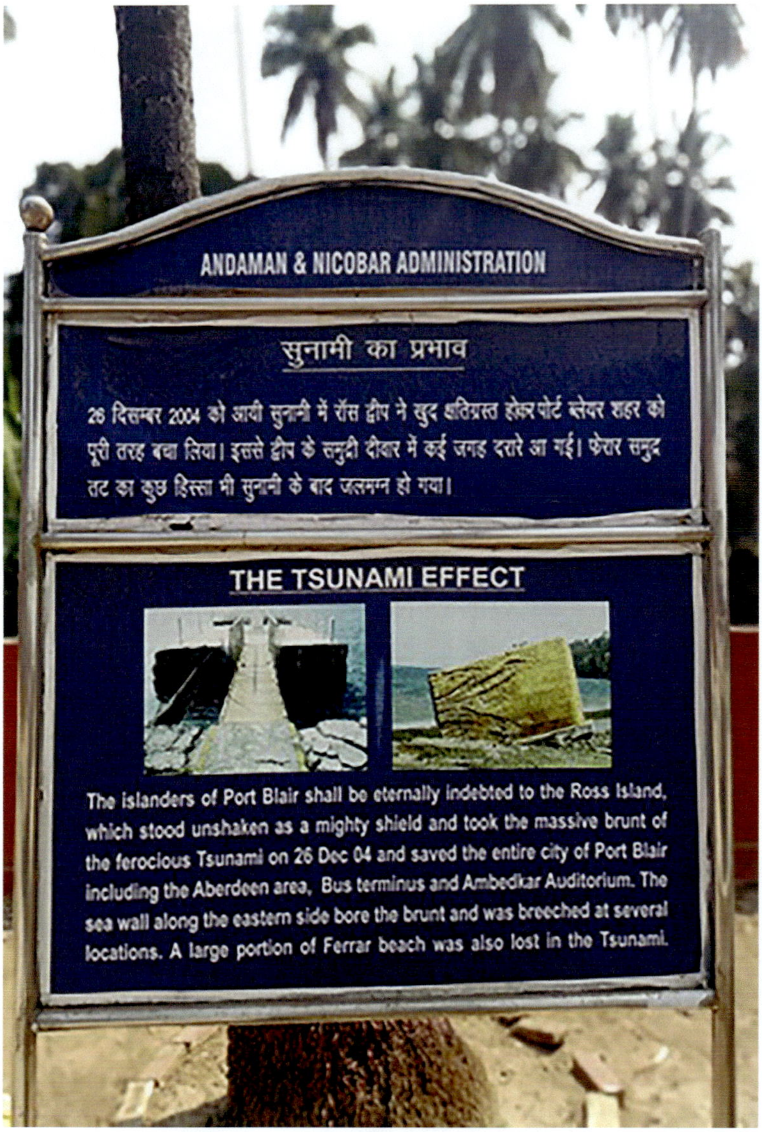

FIGURE 3.10 On the Ross islands in Andaman we find a board describing how it had protected the Port Blair.

tsunamis for the Indian region (Purnachandra Rao, 2007). The differential stress conditions obtained from b value and the material heterogeneity using fractal dimension were studied by Srivastava et al. (2015) for the Andaman-Sumatra subduction zone. As seen in Figure 3.14b, the seismicity along the subduction zone is thrust type, whereas away from the subduction zone strike slip dominates. Hence, we see that the earthquake of April 11, 2012 is strike-slip in nature and did not generate a tsunami.

Curray et al. (1982) studied the 5000-km-long and linear small seamount chain, which is buried beneath the sediments of the Bengal fan from 34.34°S to 10°N, in the Indian Ocean. It is known as the Ninety East Ridge (90° Ridge, also referred to as NER). NER was formed by the Kerguelen hot spot volcanism. The hot spot was located beneath the Indian plate during the late Cretaceous and early Cenozoic complex zone of deformation in the Indian Ocean (Pringle et al., 2008). This

TABLE 3.2
Shelf width at different locations along the west coast of India

S.No	Locations	Longitude (DD)	Latitude (DD)	Shelf width (km)
1	Dwarka	68.052	22.440	132
2	Mumbai	71.375	18.991	278
3	Ratnagiri	72.675	16.941	139
4	Panaji (Goa)	73.475	15.758	95
5	Bhatkal	74.358	13.975	132
6	Udipi	74.358	13.325	118
7	Mangalore	74.608	12.541	84
8	Thalessary	75.375	11.591	98
9	Calicut	75.125	11.341	70
10	Cochin	76.025	10.09	61
11	Thiruvanthapuram	76.691	8.525	59

ridge has a complex morphology along its entire length with an average depth of about 2500 m, and the width varies from 150 to 250 m. The northern part of the ridge is narrower and deeper than the southern part, as shown by Clater and Fisher (1974). To the west of this significant feature in the eastern Indian Ocean is the central Indian basin while the Wharton basin lies to its east. Most of the seismic activity along the NER is characterized by strike-slip earthquakes. The flanks on either side of the ridge exhibit a consistent topographic gradient. The eastern flank is about 1 km deeper than the western side. A similar study to a tsunami passing over Murray Ridge has been carried out for a tsunami going over NER with and without bathymetry, and the effect of NER on the tsunami directivity, arrival times and wave heights was quantified by Swapna (2018) (Table 3.3).

The December 26, 2004 Sumatra earthquake ruptured the Andaman-Sumatra subduction zone from 3°N to 14°N latitudes, about 1400 km in length, and the orientation of the fault is not uniform. Ioualalen et al. (2007) divided the rupture zone into five segments. Considering different slip distributions for the five segments we have modeled initial deformation at these five segments and then combined all the outputs into a single source and considered this to be an initial wave to model

Three profiles, L, M, and N, across the NER with three gauge locations on each profile are considered (Figure. 3.15). Maximum wave height plots along the gauge locations L, M, and N profiles are given in Figure 3.16. The directivity for two scenarios, I and II, with and without the NER, is computed, and the results are plotted in Figure 3.17.

The initial displacement at the source for scenario I is around 6 m. It is clearly observed that the waves were amplified at L2, M2, and N2 locations in two scenarios (Figure 3.16). The directivity of tsunami wave propagation after reaching the NER deviates, and increased wave heights are observed after crossing the NER. Without NER, the directivity is directed towards the open ocean with decreased wave heights. With NER, wave heights are greater along the Indian coast than without the NER system along the east coast of India (Swapna, 2018). The findings of Titov et al. (2005), Groesen et al. (2008), and Andrei (2009) also suggest the effect of ridges in the ocean affecting the directivity and also the focusing of the waves. Usually, the directivity of the wave propagation is perpendicular to the orientation of the fault (strike angle). Figure 3.17 shows the directivity of tsunami wave propagation for the 1400-km ruptured zone. The wave propagated transoceanic regions and reached India's east and west coast. The NER guided the wave energy towards the south. The waves, after reaching the Chagos-Laccadive ridge, got reflected. Because of the presence of these ridges, the extensive coastline of Africa and Madagascar was saved from a major tsunami threat.

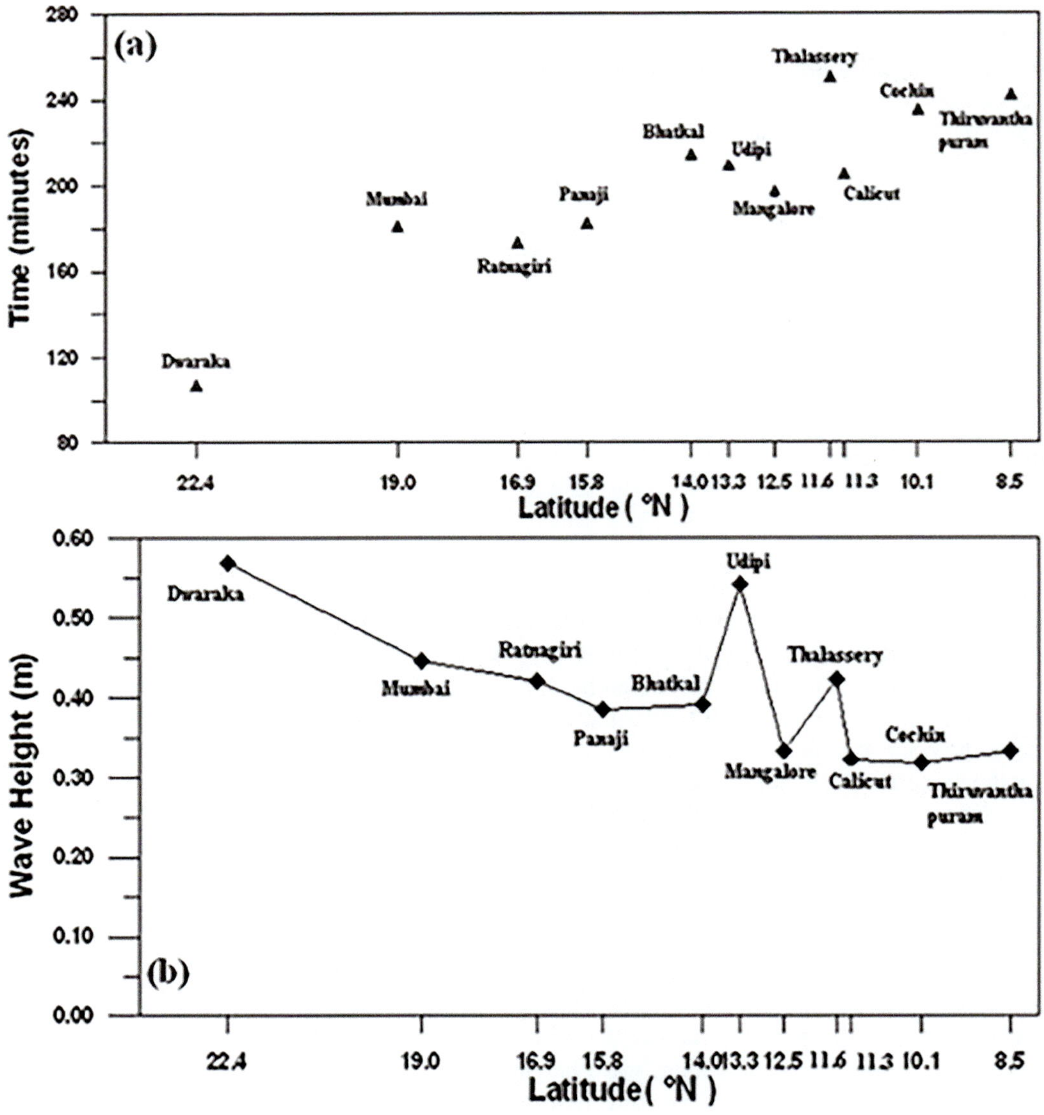

FIGURE 3.11 Tsunami (a) arrival times and (b) tsunami wave heights at different locations along the west coast of India.

3.7 SRI LANKA SHIELDING THE SOUTHERN TIP OF INDIA

Kudankulam is located at the bottom tip of India, and it was also affected due to the December 26, 2004 Sumatra tsunami. Sri Lanka's geographical position as an island located to the south of India served as a protective barrier, mitigating the tsunami's direct effect on the coastal regions of Tamil Nadu. The height of the tsunami run-up was diminished in areas that were sheltered by Sri Lanka in contrast to those that were directly exposed to the open ocean. Findings suggest that the secondary waves were amplified on the Kudankulam coast. This is because due to the presence of Sri Lanka, the tsunami wave were reflected between India and Sri Lanka, thereby amplifying the secondary wave. The run-up heights at different locations ranged from 2 m to 4 m Krishna (2013). The tsunami affected both Sri Lanka and India, although the degree of impact was different. The southern coast

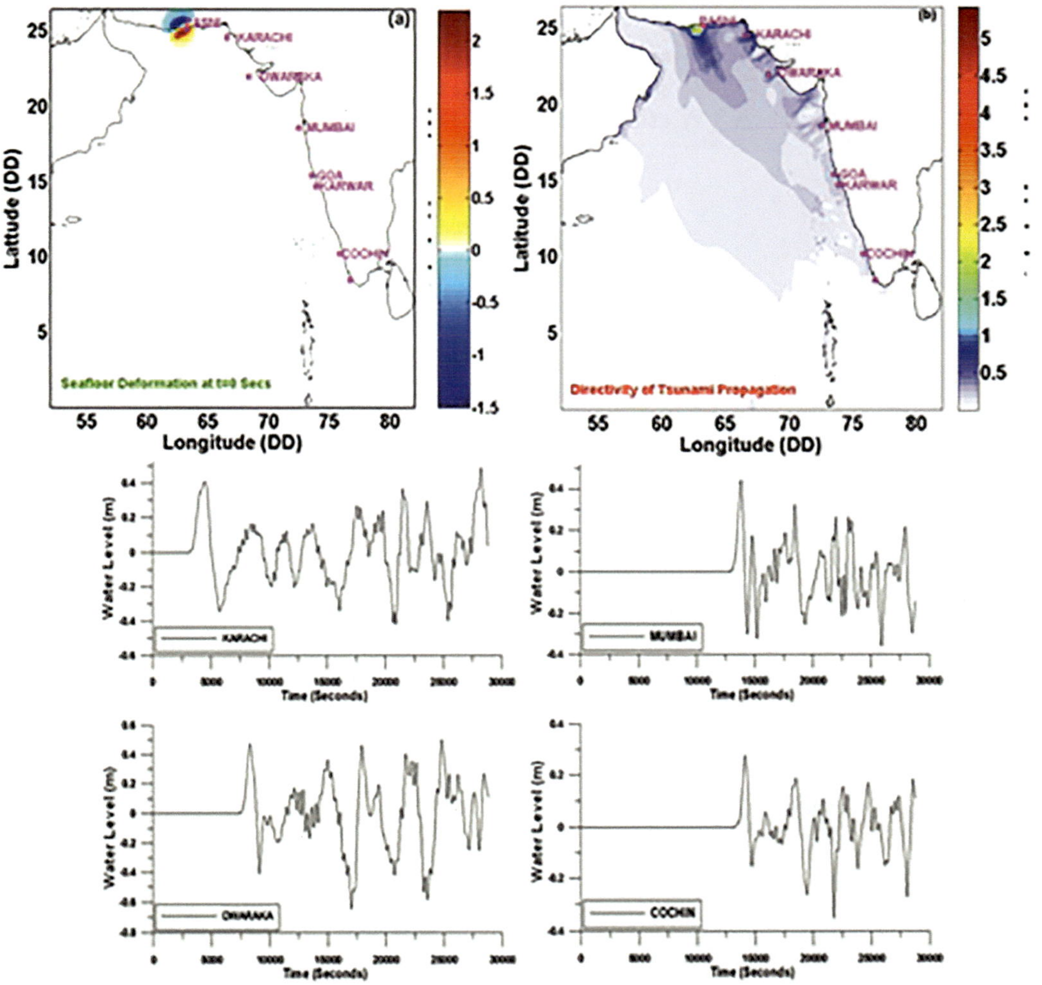

FIGURE 3.12 (a) Initial deformation at the source, (b) directivity of tsunami, and (c) the estimated run-up heights near Mumbai due to 1945 Makran tsunami.

of Sri Lanka suffered severe damage, whereas the southern coast of India, especially the regions protected by Sri Lanka, encountered milder effects (Figure 3.17)

3.8 SUMMARY

Tsunamis represent a natural hazard, capable of causing extensive damage and loss of life along coastlines. Effective tsunami risk mitigation requires a multi-faceted approach. Investment in early warning systems and public education is crucial. Vulnerability mapping, inundation modeling, and run-up and arrival time analysis are essential tools for understanding and predicting the impact of tsunamis. Shoreline change monitoring can be used to assess coastal resilience and to adapt infrastructure and evacuation plans accordingly.

Tsunami propagation from the MSZ across the MR was quantified. Without the MR, we observe that the waves travel faster and reach the coast earlier without amplification. However, in the presence of the MR, the tsunami waves slow down and are amplified above the MR and propagate

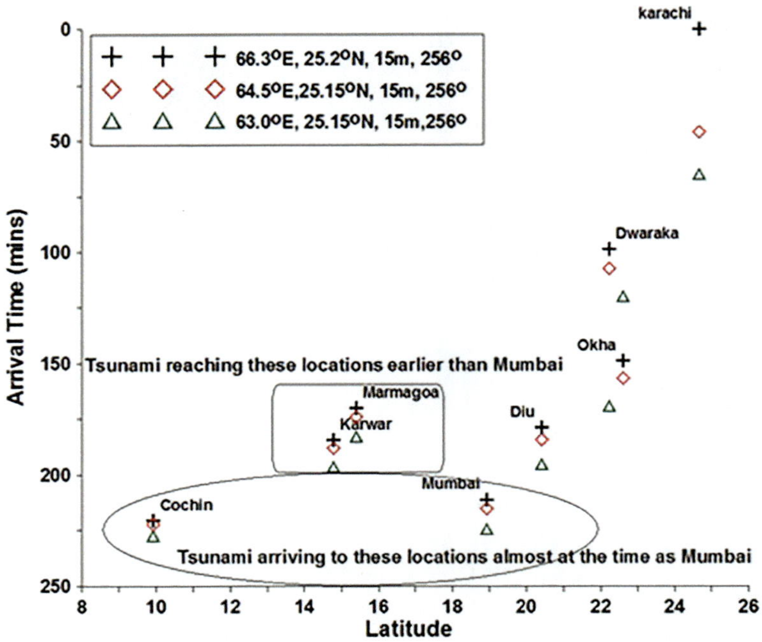

FIGURE 3.13 Tsunami wave arrival times to different locations along the west coast of India due to possible scenarios from the Eastern Makran.

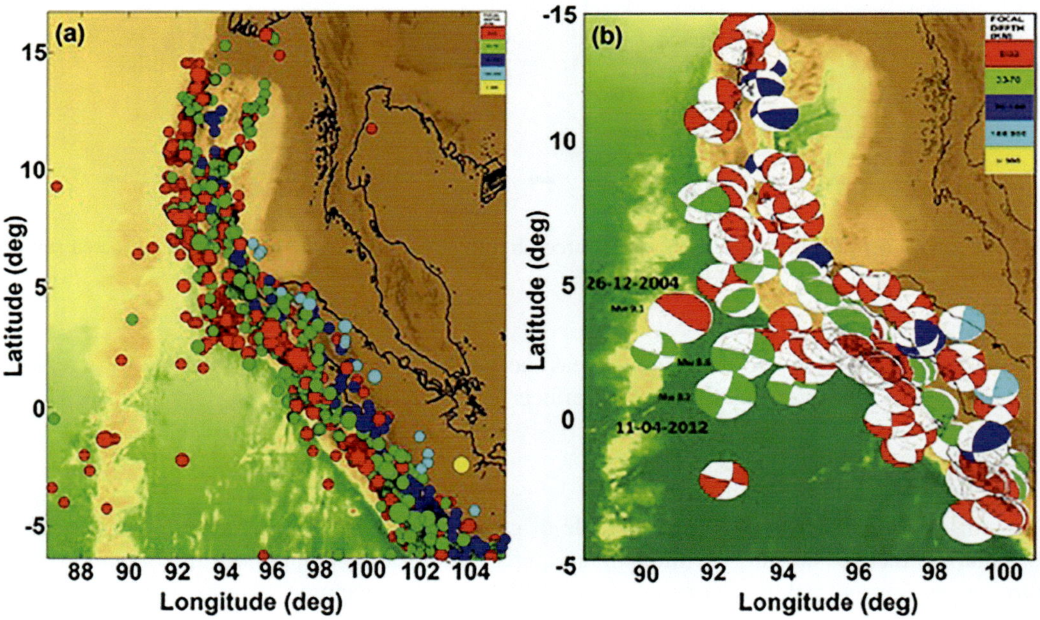

FIGURE 3.14 (a) Seismicity (1990–2009) along the ASSZ with magnitude >4. and (b) focal mechanisms (1976–2014) of earthquake magnitude > 6.

TABLE 3.3
Fault parameters used for tsunami simulations from the Andaman-Sumatra subduction zone

	Scenario-I Real case of 2004 Tsunami				
(Ioualalen et al 2007)	**Segment-1**	**Segment-2**	**Segment-3**	**Segment-4**	**Segment-5**
Coordinates	95.10°E To 2.50°N	93.90°E To 4.33°N	93.41°E To 5.80°N	92.10°E To 9.10°N	92.0°E To 10.5°N
Length (km)	220	150	390	150	350
Width (km)	130	130	120	95	95
Depth (km)	25	25	25	25	25
Dip (degrees)	12	12	12	12	12
Strike (degrees)	330.0	338	338	356	10
Rake (degrees)	90	90	90	90	90
Slip (m)	15	15	06	06	06
Moment Magnitude	9.4				

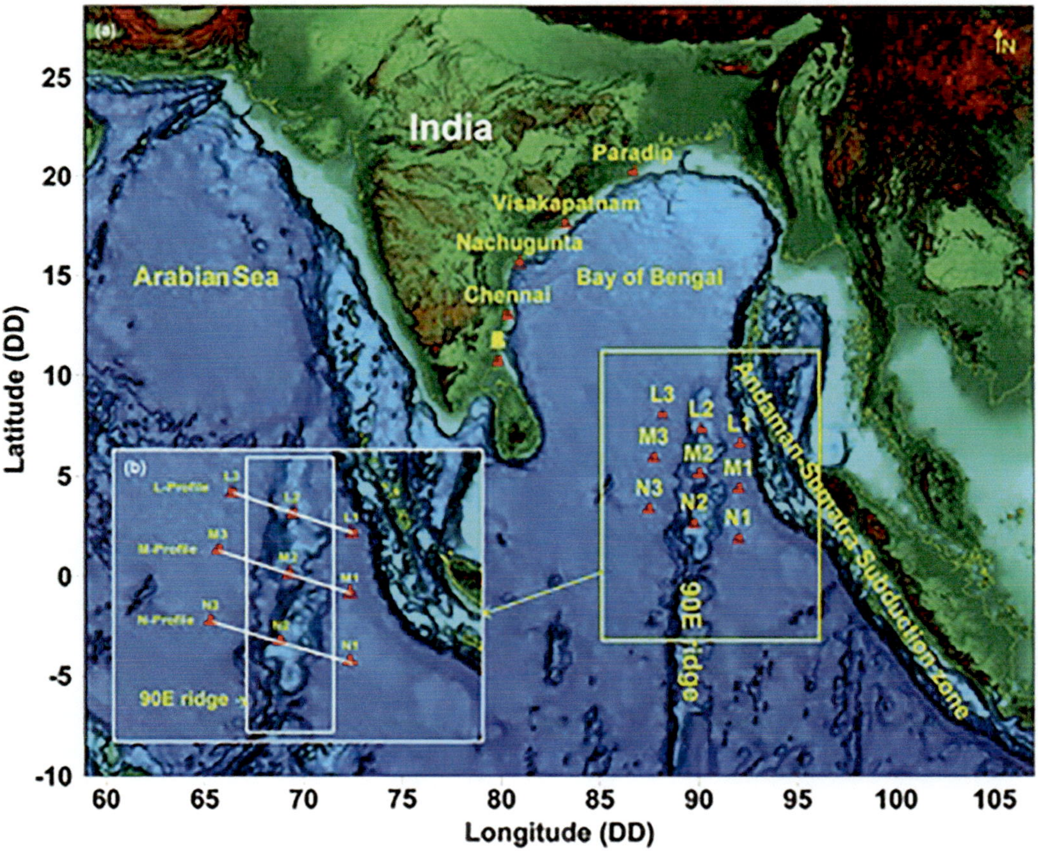

FIGURE 3.15 (a) Tide gauge locations were chosen for arrival time and wave height computation in A-grid. (b) Gauge locations taken along L, M, and N profiles across the NER ridge.

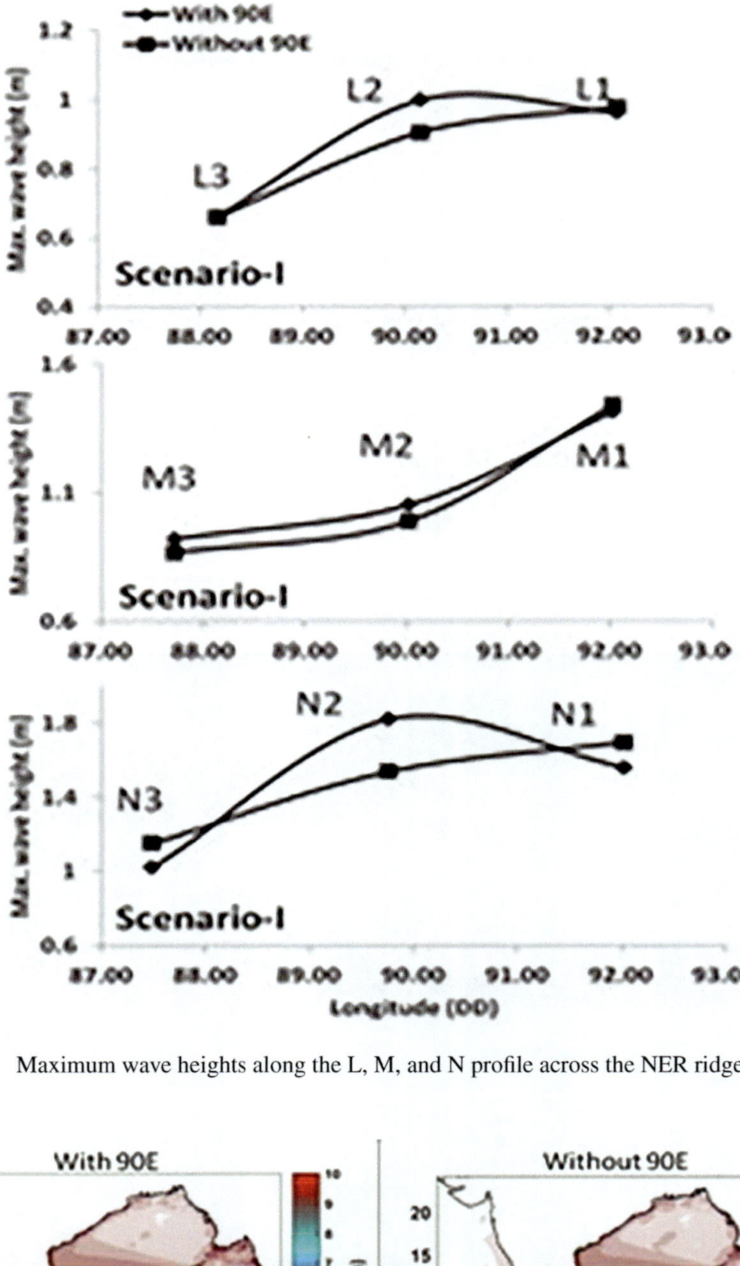

FIGURE 3.16 Maximum wave heights along the L, M, and N profile across the NER ridge.

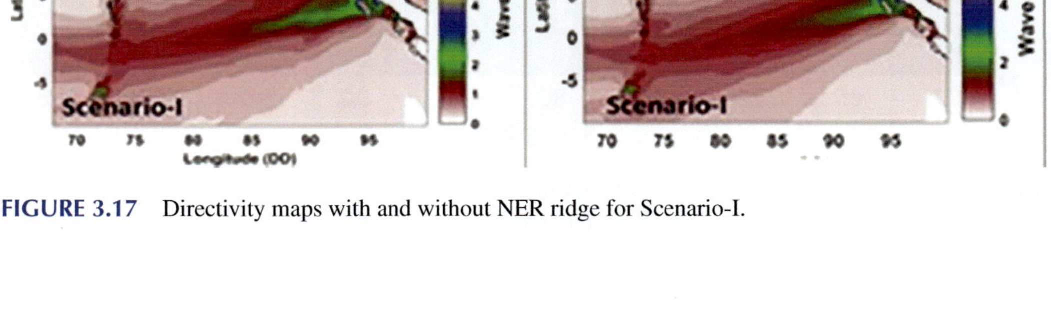

FIGURE 3.17 Directivity maps with and without NER ridge for Scenario-I.

through the ocean, causing the late arrival of tsunami waves along the west coast of India and with increased wave heights.

In the presence and absence of Lakshadweep Ridge's bathymetry, the tsunami along India's southwest coastal part was studied, considering the 1945 Makran tsunami at the subduction zone of Makran. To observe the brunt alongside the western coastal part of India, a few locations were considered across the islands of Lakshadweep, and the heights of the waves were quantified at various selected locations. Due to the presence of the Lakshadweep islands, the tsunami waves are amplified, and there was an increase in the arrival time of tsunami waves at the southern part of India's western coast. However, without considering the Lakshadweep islands' bathymetry, the arrival times of tsunamis and the amplitudes of waves close to Mangalore on India's coastal mainland have decreased. The Islands of Lakshadweep act as a wave barrier for India's southwestern coastal region. Because of an extensive shelf width along the western coastal part of India, only a minimal extent of inundation is observed. The changing shelf width along the west coast causes a change in tsunami arrival times.

In another study, Tsunami wave heights along the three profiles across the NER are considered, and wave heights and arrival times were quantified. With NER, the tsunami waves travelled slowly due to the shallow depth of NER and thus took longer to reach the coast with amplified wave height. Without NER, the tsunami waves reach the coast faster. The shielding effect of Sri Lanka on the southern tip of India was also quantified.

REFERENCES

Ambraseys, N. N., & Melville, C. P. (1982). *A history of Persian earthquakes* (p. 219). Cambridge University Press, London.

Buchward, V. T. (1968). Long waves on oceanic ridges. *Proceedings of the Royal Society of London. Series A, 308*, 343–354, Printed in Great Britain.

Byrne, D. E., Sykes, L. R. & Davis, D. M., (1992). Great thrust earthquakes and seismic slip along the plate boundary of the Makran subduction zone. *Journal of Geophysical Research, 97*, 449–478. doi: 10.1029/91JB02165

Chamoli, A., Swaroopa Rani, V., Srivastava, K., Srinagesh, D., & Dimri, V. P. (2010). Wavelet analysis of the seismograms for tsunami warning. *Nonlinear Processes in Geophysics, 17*, 569–574. doi: 10.5194/npg-17-569-2010

Curray, J, R., Emmel, F, J., Moore, D. G., & Raitt, R. W. (1982). Structure, tectonics and geological history of the northeastern Indian Ocean. In A. E. Nairn & F. G. Stehli (Eds.), *The ocean basins and margins* (Vol. 6, pp. 399–450). Plenum Press, New York.

Edwards, R.A., Minshull, T.A., & White, R. S. (2000). Extension across the Indian–Arabian plate boundary: The Murray Ridge. *Geophysical Journal International, 142*, 461–477. doi: 10.1046/j.1365-246x.2000.00163

Farhoudi, G., & D. E. Karig. (1977). The Makran of Iran and Pakistan as an active Arc System. *Geology, 5*(11), 664–668.

Gaedicke, C., Schlijter, H.U., Roeser, H.A., Prexl, A., Schreckenberger, B., Meyer H., Reichert, C., Clift, P., & Amjad, S. (2002). Origin of the northern Indus Fan and Murray Ridge, northern Arabian Sea: Interpretation from seismic and magnetic imaging, *Tectonophysics, 355*, 127–143. doi: 10.1016/S0040-1951(02)00137-3

Groesen, E. V., Adytia, D., & Andonowati. (2008). Near-coast tsunami wave guiding: Phenomenon and simulations. *Natural Hazards and Earth System Sciences, 8*, 175–185. doi: 10.5194/nhess-8-175-2008

Heck, N. H. (1947). List of seismic sea wave. *Bulletin of the Seismological Society of America, 37*(4), 269–286.

Heidarzadeh, M., & Kijko, A. (2011). A probabilistic tsunami hazard assessment for the Makran subduction zone at the northwestern Indian Ocean. *Natural Hazards, 56*, 577–593. doi: 10.1007/s11069-010-9574-x

Heidarzadeh, M., Pirooz, M. D., Zaker, N. H., Yalciner, A. C., Mokhtari, M., & Esmaeily, A. (2008). Historical tsunami in the Makran Subduction Zone off the southern coasts of Iran and Pakistan and results of numerical modeling. *Ocean Engineering, 35*, 774–786. doi: 10.1016/j.oceaneng.2007.10.005

Ioualalen, M., Asavanant, J., Kaewbanjak, N., Grilli, S. T., Kirby, J. T., & Watts, P. (2007). Modeling the 26 December 2004 Indian Ocean Tsunami: Case study of impact in Thailand. *Journal of Geophysical Research, 112*, C07024, doi: 10.1029/2006JC003850

Koppa, C., Fruehn, J., Flueh, E. R., Reichert, C., Kukowski, N., Bialas, J., & Klaeschen, D. (2000). Structure of the Makran subduction zone from wide-angle and reflection seismic data. *Tectonophysics, 329*, 171–191.

Krishna, K. (2013). *Inundation studies for a few nuclear power plants and other major cities along the east and west coast of India due to tsunamigenic earthquakes in the Indian Ocean.* PhD Thesis, Osmania University, Hyderabad, India.

Lay, T., Kanamori, H., Ammon, C. J., Nettles, M., Ward, S. N., Aster, R. C., ... & Sipkin, S. (2005). The great Sumatra-Andaman earthquake of 26 December 2004. *Science, 308*(5725), 1127–1133. doi: 10.1126/science.1112250

Marchuk, A. G. (2009). Tsunami wave propagation along waveguides. *Science of Tsunami Hazards , 28*(5), 283–302.

Minshull, T. A., White, R. S., Barton, P. J., & Collier, J. C. (1992). Deformation at plate boundaries around the Gulf of Oman. *Marine Geology, 104*, 265–277.

Mita Rajaram, Anand, S. P., & Majumdar, T. J. (2009). Structure and tectonics of the Indian offshore region from satellite-derived geopotential data. *Proceedings of the second swarm international science meeting held at GFZ*, Potsdam, Germany, 24 to 26 June.

Mofjeld, H. O., Symons, C. M., Lonsdale, P., González, F. I., & Titov, V. V. (2004). Tsunami scattering and earthquake faults in the deep Pacific Ocean. *Oceanography, 17*(1), 38–46. doi: 10.5670/oceanog.2004.65

Murty, T.S., & Bapat, A. (1999). Tsunamis on the coastlines of India. *Science of Tsunami Hazards, 17*(3), 167–172.

Murty, T. S., & Rafiq, M. (1991). A tentative list of tsunamis in the marginal sea of the North Indian Ocean. *Natural Hazards, 4*, 81–83.

Neetu, S., Suresh, I., Shankar, R., Nagarajan, B., Sharma, R., Shenoi, S. S. C., Unnikrishnan, A. S., & Sundar, D. (2011). Trapped waves of the 27 November 1945 Makran tsunami: Observations and numerical modeling. *Natural Hazards, 59*, 1609–1618. doi 10.1007/s11069-011-9854-0.

Okal, E. A. (2015). The quest for wisdom: lessons from 17 Tsunamis, 2004–2014. *Philosophical Transactions of the Royal Society A: Mathematical, Physical and Engineering Sciences, 373*(2053), 20140370.

Page, W. D., Alt, J. N., Cluff, S., & Plafker, G. (1979). Evidence for the recurrence of large magnitude earthquakes along the Makran Coast of Iran and Pakistan. *Tectonophysics, 52*, 533–547. doi: 10.1016/0040-1951(79)90269-5

Park, J., Song, T.-R. A ., Tromp, J., Okal, E., Stein, S., Roult, G., Clevede, E., Laske, G., Kanamori, H., Davis, P., Berger, J., Braitenberg, C., Van Camp, M., Lei, X., Sun, H., Xu, H., & Rossat, S.(2005). Earth's free oscillations excited by the 26 December 2004 Sumatra-Andaman earthquake. *Science, 308*, 1139–1144. doi: 10.1126/science.1112305

Pendse, C. G. (1948). The Makran earthquake of the 28 th of November, 1945. *Indian Meterological Department Science Notes, 10*, 141–145.

Pringle, M. S., Frey, F. A., & Mervine, E. M. (2008). A simple linear age progression for the Ninetyeast Ridge, Indian Ocean: New constraints on Indian plate motion and hot spot dynamics. *Eos, Transactions American Geophysical Union, 89*(53), Fall Meet. Suppl., Abstract T54B-03.

Quittmeyer, R. C. K., & Jacob, H. (1979). Historical and modern seismicity of Pakistan, Afghanistan, Northwestern India, and Southeastern Iran. *Bulletin of the Seismological Society of America, 69*(3), 773–823.

Rajendran, C. P., Ramanamurthy, M. V., Reddy, N. T., & Rajendran, K. (2008). Hazard implications of the late arrival of the 1945 Makran tsunami. *Current Science, 95*, 1739–1743.

Rao, P. (2007). Characterization of potential tsunamigenic earthquake source zones in the Indian Ocean. In A. Kundu (Ed.), *Tsunami and nonlinear waves* (pp. 285–312). Springer, Berlin, Heidelberg.

Saha, S. (2022). *Mathematical data modeling of tsunamis from Makran Subduction Zone and its impact on Lakshadweep islands and West coast of India.* PhD Thesis, Andhra University, Vishakhapatnam, India.

Saha, S., & Srivastava, K. (2016). Effect of tsunamis from Makran subduction zone and its impact on the Androth Island of Lakshadweep. *Natural Hazards, 82*(3), 1755–1766. doi: 10.1007/s11069-016-2267-3

Saha, S., & Srivastava, K. (2019). Tsunami Simulation along the South West Coast of India: Lakshadweep Islands acting as wave-barricade. *Indian Journal of Marine Sciences, V48*(10), 1640–1647.

Satake, K. (1988). Effect of bathymetry on tsunami propagation: Application of ray tracing to tsunamis. *PAGEOPH, 126*, 27–36. doi: 10.1007/BF00876912

Schluter, H. U., Prexl, A., Gaedicke, C., Roeser, H. C., Reichert, C., Meyer, H., & von Daniels, C. (2002). The Makran Accretionary Wedge: Sediment thickness and ages and the origin of mud volcanoes. *Marine Geology, 185*, 219–232. doi: 10.1016/S0025-3227(02)00192-5

Sclater, J. G., & Fisher, R. L. (1974). Evolution of the east: Central Indian Ocean, with emphasis on the tectonic setting of the Ninetyeast Ridge. *Geological Society of America Bulletin, 85*(5), 683–702.

Srivastava, K., Dimri, V. P., Rani, V. S., Kumar, R. K., & Narain, L. (2011). Would Makran tsunami skip Mumbai, India? No it would reach 8 minutes later than Ratnagiri. *Indian Journal of Marine Sciences, 40*(5), 620–623.

Srivastava, K., Rani, S., & Srinagesh, D. (2015). A review of b-value imaging and fractal dimension studies in the Andaman Sumatra subduction. *Natural Hazards, 77*, 97–107. doi: 10.1007/s11069-014-1143-2

Stein, S., & Okal, E. (2005). Speed and size of the Sumatra earthquake. *Nature, 434*, 581–582. doi: 10.1038/434581a

Swapna, M. (2018). *Tsunami modeling and sensitivity studies over ridges in the Indian Ocean*. PhD Thesis, Osmania University, Hyderabad, Inida.

Swapna, M., & Kirti, S. (2014). Effect of Murray ridge on the Tsunami propagation from Makran Subduction zone. *Geophysical Journal International, 199*, 1430–1441. doi: 10.1093/gji/ggu336

Swaroopa, R., Kirti, S., Srinagesh, D., & Dimri, V. P. (2011). Spatial and temporal variation of b-value and fractal dimension in the Makran Subduction Zone. *Marine Geodesy, 34*(1), 77–82. doi: 10.1080/01490419.2011.547804

Swaroopa, R. V. (2011). *Tsunami modeling and hazard estimation along the east and west coast of India due to tsunamigenic earthquakes in the Indian ocean*. PhD Thesis, Osmania University, Hyderabad, India.

Titov, V. V., Rabinovich, A. B., Mofjeld, H. O., Thomson, R. E. & Gonzalez, F. I. (2005). The global reach of the 26 December 2004 Sumatra tsunami, *Science, 309*, 2045–2048. doi: 10.1126/science.1114576

White, R. S. (1984). Active and passive boundaries around the Gulf of Oman, northwest Indian Ocean. *Deep Sea Research, 31*, 731–745.

Yeh, H., Titov, V., Gusiakov, V., Pelinovsky, E., Khramushin, V., & Kaistrenko, V. (1995). The 1994 Shikotan earthquake tsunamis. *PAGEOPH, 144*(3/4), 855–874.

4 Tsunami-induced Forces on Coastal Regions of India

4.1 INTRODUCTION

Scientists have been working hard to study the effect of tsunami waves on collapsed buildings to improve our understanding of the different forces acting during a tsunami. A large body of research (Kumaraguru et al., 2005; Kunkel et al., 2006; Mizutani et al., 2001) has been devoted to learning about the way conventional coastal protection systems—seawalls, coral reefs, and breakwaters—react to the hydrodynamic force from tsunamis. For example, Kumaraguru et al. (2005) investigated the influence of tsunamis on coral reefs, contributing to information that helps measure the reduction in tsunami impacts when conducting modeling tasks (Kunkel et al., 2006). Mizutani et al. (2001) also carried out wave force measurements that affect coastal protection structures such as seawalls and breakwaters.

In spite of this, not much research has gone into investigating how tsunamis affect other structures like bridges and coastline structures. Of particular note is a study by Azadbakht et al. (2014) aimed at estimating the forces caused by tsunamis on bridge structures using simulations, distinguishing between the first impact and the full wave inundation phases. Moreover, examining the patterns of failure in impacted buildings can yield valuable information on how to develop correct design equations for tsunami-induced hydrodynamic loads—a technique used by Chock et al. (2012) in their research on structural damage caused by the Tohoku tsunami.

The extensive damage resulting from the Indian Ocean tsunami on December 26, 2004, which heavily affected coastal communities in nations such as Thailand, Sri Lanka, Indonesia, and India, emphasized the necessity of assessing the structural resilience of coastal structures. Nistor et al. (2009) analyzed the forces produced by tsunami-induced hydraulic bores and the impact of debris, including example calculations of tsunami loads on model structures.

In addition, experimental studies have also been carried out to predict the hydrodynamic forces of tsunamis on structures located along coastlines. One such experiment by Suzuki et al. (2014) focused on the Tohoku-Pacific Ocean earthquake of 2011, which caused extensive damage to submerged bridges. Conventional structural design did not expect tsunami waves to wash away bridges' components, such as girders, and procedures to predict such forces were previously lacking. To deal with this, Suzuki et al. (2014) conducted physical modeling and 2D and 3D numerical simulations to examine tsunami behavior and resulting forces on bridge structures.

Surges or bores have frequently been observed in tsunami-prone areas. On several occasions, these bores have been very destructive. Hence a study on forces due to bores has become very significant. Robertson et al. (2013) conducted experiments to compute forces due to tsunami bores on coastal structures. He proposed a design equation for forces due to bores for numerous bore heights and vertical water depths.

DOI: 10.1201/9781003784500-4

The huge impact of tsunami-induced bores on coastal structures has been documented well during the recent crisis in the Indian Ocean 2004 tsunami, Chile 2010 tsunami, and Japan 2011 tsunami. Based on surveys from site visits to these regions, Palermo et al. (2012) conducted experiments to investigate further tsunami loading on coastal structures. Nouri et al. (2010) worked on an extensive study of the hydraulic bore–structure interaction, executed using an experimental apparatus creating bores, aimed to give a better comprehension, an estimate of the time history of forces both quantitatively and qualitatively, and forces due to a hydraulic bore, like the tsunami-induced coastal flooding, on structures.

Similarly, Cross (1967) presented experimental data for the surge tip shape and the forces produced on impact. He also gave an equation to measure surge forces in different conditions. A comparative study on forces due to surges and bores has also been done by Nistor et al. (2009).

Though people have extensively worked on tsunami-induced forces on structures, a good knowledge of tsunami wave behavior is required to use these studies when engineering structures. Hence, there is a great need for civil engineers to have the data on wave characteristics and forces to engineer better structures that can sustain less damage due to tsunamis.

To calculate the forces, a local survey of coastland with run-up heights and wave characteristics is needed. The above experimental studies can be used to study the wave characteristics at different locations and calculate forces.

4.2 TYPES OF FORCES

4.2.1 Hydrostatic Force

The resultant force caused by the pressure loading of a liquid on the submerged surfaces is called the Hydrostatic force. The Hydrostatic pressure at a depth d can be defined as

$$P_H = \rho g d$$

where ρ is the water density and g is the gravitational acceleration.

This force is considered while engineering structures in oceans and seas like oil rigs.

4.2.2 Hydrodynamic Force

Hydrodynamic force, also called the drag force, arises from water particle velocities and acceleration. These forces can vary (when generated by waves) or can be constant (when generated by steady currents) and bring about a dynamic load on the structure in their path.

These forces are the main reason for damage to coastal structures.

4.2.3 Impact Force

In a tsunami event, a lot of debris (wood pieces, metal, furniture, boats, etc.) is pulled by the waves. Being hit by debris causes much more damage compared to the water wave. The impact force is the force of the debris on coastal structures. Since calculating this force is completely probabilistic, we are not taking it into account.

4.3 TYPES OF WAVES

4.3.1 Wave Breaking Criteria (Deo, 2013)

Depending on their run-up heights and seabed slope, waves break in a depth ranging from about 0.8 to 2.4 times their height. The coastal structures are subjected to the action of three different waves

TABLE 4.1
Wave breaking criteria

Non-breaking waves	Breaking waves	Broken waves
D > 2.4H	0.8H< D <2.4H	D <0.8H

depending on the water depth at their location and wave heights. The depth to height relation for the wave breaking is given in Table 4.1.

4.3.2 Non-Breaking Wave Force

On a smooth-faced vertical wall, the non-breaking wave would be completely reflected. Hence, standing waves will be formed. The pressure at the depth z is given by (Sainflou, 1928)

$$p = \gamma h \ \text{coskx} \ \cos \omega t \frac{\text{coshk}(d+z)}{\text{coshkd}} - \gamma z$$

where γz is static and $\gamma H \ \text{coskx} \ \cos \omega t \frac{\text{coshk}(d+z)}{\text{coshkd}}$ is the dynamic part.

Here we assume linear theory to be valid.

Choosing x = 0, cos(kx) =2.

Here ρ is the water density, g is the gravitational acceleration, d is below mean sea level water depth, b is the length of the section of the dam, v is the incoming wave velocity, and h is the wave height above mean sea level.

Here γ=ρg, H is the difference in amplitude between the maximum crest and maximum trough, k is the wavenumber of the tsunami wave, ω=2π/T, and T is the time period of the wave.

4.3.3 Breaking Wave Force

The maximum dynamic pressure (Deo, 2013) due to wave breaking is given by:

$$P_m = 101 \rho g \frac{H_b}{\lambda_D} \frac{d_s}{dl} \left(dl + d_s\right)$$

Here λ_D = wavelength at depth D, dl= depth at one wavelength away from the structure. We can observe a parabolic variation over the wall.

4.3.4 Broken Wave Force

The broken wave mass travels towards the shore with the same velocity as those before breaking. Broken waves form surges. The total surge force per unit width is

$$F / w = 4.5 \rho g h^2$$

here ρ is the water density, g is the gravitational acceleration, and h is the height of the wave.

4.4 STUDY AREA: PUDUCHERRY

We have considered Puducherry as the area of interest specifically due to the Beach road which was not impacted during the 2004 Indian Ocean tsunami. The structures along this road were not damaged. Also the presence of other beaches with different slopes can help us understand the inundation and forces on structures in a better way.

The area of study, Puducherry, lies on the eastern coast of India, between latitudes 22.05°N to 22.75°N and longitudes 79.87°E to 79.79°E. The coastal area is at a mean elevation near sea level and has many coastal inlets, commonly known as backwaters. The topography of Puducherry's coast is flat and gently sloping, making it more vulnerable to flooding. Murali et al. (2013) performed a vulnerability analysis of this seashore, and there has been a lot of research regarding the effects of tsunamis by Murthy et al. (2011) and Anandabhaskaran (2017).

Narayanan et al. (2015), through the examination of depth profiles along the shore, determined that there is an extensive underwater feature called the Puducherry Canyon. The gradient of the seafloor within this region is gentle, varying from 0 to 25 m depth along a distance of 22 km offshore. The maximum depth, at 300 m, occurs around 22 km off the coast, near the Puducherry Canyon.

The 2004 Indian Ocean Tsunami struck many South Asian countries, causing great damage along the Andaman-Nicobar and Indian coasts. Though the major devastation was along the Nagapattinam region, Puducherry coasts experienced significant impact. Elumalai et al. (2006) undertook a damage assessment using Remote Sensing and GIS, concerning natural and physical resources, in parts of Cuddalore and Puducherry coasts. The structural damage was studied village-wise and categorized as maximum, moderate, and least affected areas. As far as the natural resources are concerned, the changes in geomorphology, land use/ land cover, soil, and water have been studied. Based on the same, the risk maps were generated in the GIS environment.

Even though India has developed tsunami warning systems since the 2004 disaster, they can provide data only on the likely coasts to be struck by the tsunami and not on the specific areas that would be flooded, which is the need of the hour. There has been a study conducted by Ramasamy (2014) in some areas of Tamil Nadu and Puducherry coast embracing the Indian east coast's southern part. The research was conducted by the reconstruction of tsunami surge paths within the coastal terrain, from the damage pattern, type, and intensity of damages inflicted by the tsunami (2004); and thereby deriving the contribution of coastal systems to the movement of tsunamis on to the land, such that these are included in the tsunami warning systems for effective forewarning, including areas susceptible for inundation. The research showed that the directions of tsunami surge on the coastal terrain were largely determined by the geometry of the coastal zone or the geological formations and the tectonic processes that supplied such geometry, the vegetative cover, and also the human activities in the form of structures and the mode of their arrangement.

Sheth et al. (2006) has surveyed the damage due to the 2004 disaster across mainland India. It has been observed that the behavior of tsunami waves on the mainland is affected by the geographic and topological features of the coast. Extensive damage has been suffered by the houses and structures along the coastal region, besides roads and bridges. Destruction of the structures was both by direct force from tsunami waves, and drag force by the receding waves. A study on the affected structures showed that many of them were non-engineered and poorly built houses of the fishing community. It has been observed that about 20,000 dwelling units were lost or damaged in Puducherry itself. Among the damaged structures were a 2.4-m-high wall, two adjacent arches of Arasalar Bridge, which collapsed, and a helipad washed away on the southern side of a backwater, along with a wooden army jetty. This survey proves the significant damage to the coastal economy due to the tsunami. Hence, there is a need for calculating tsunami-induced forces along coastal regions.

In this study, an attempt has been made to calculate forces along the three coastal regions of Puducherry: Auroville Beach, Beach Road, and Paradise Beach, and also provide a vulnerability map for the entire Puducherry coast.

4.4.1 Coastal Survey of Puducherry

4.4.1.1 Topographic, Bathymetric Data

The Puducherry coast has been separated into three important study locations: (I) Beach Road, (II) Auroville Beach, and (III) Paradise Beach. CMAP data for this area has been digitized, and field data of boulder positions along Beach Road have been incorporated to obtain the bathymetry (see Figures 4.1 and 4.2). This data was later merged with SRTM and GECO height data in order to produce the nested grid system necessary for TUNAMI N2 simulation. The distribution of the three beaches in terms of space can be seen from Figure 4.7. Certain points which were identified as Beach Road 1, 2, and 3 were utilized in estimating the tsunami force at these positions. Auroville Beach is located in the northern part, Beach Road holds the middle region, and Paradise Beach is located in the southern part of the Puducherry coastline.

From observation, Paradise Beach (Figure 4.5) has a shallower shoreline gradient than the steep gradient of Auroville Beach (Figure 4.3), leading to a larger run-up of waves in the latter. Rock boulders along the entire length of Beach Road (Figure 4.4) mark its features. All three coastal locations had their tsunami wave heights and the inundation extents computed. Figure 4.6 shows the evacuation route for the tsunami.

4.4.2 Run-up along the Puducherry Coast

In this research, the tsunami of the Indian Ocean has been simulated to analyze the run-up along the coast of Puducherry. The TUNAMI-N2 numerical code is used to model (Figure 4.8) tsunami generation and wave propagation, predict tsunami arrival times, compute tsunami run-up heights, and find the distribution of coastal inundation. A graphic representation of the directional spread of the tsunami is presented in Figure 4.9. Such directivity maps have been produced by other researchers as well, and respective run-up and inundation level analyses along different coastal segments have been performed as well (Swaroopa Rani et al., 2011).

The run-up along the entire coast of Puducherry is shown in Figure 4.10. An increase in wave heights is observed towards the southern region of the coast. Inundation is the highest around Paradise Beach and very low or no inundation is observed along Beach Road.

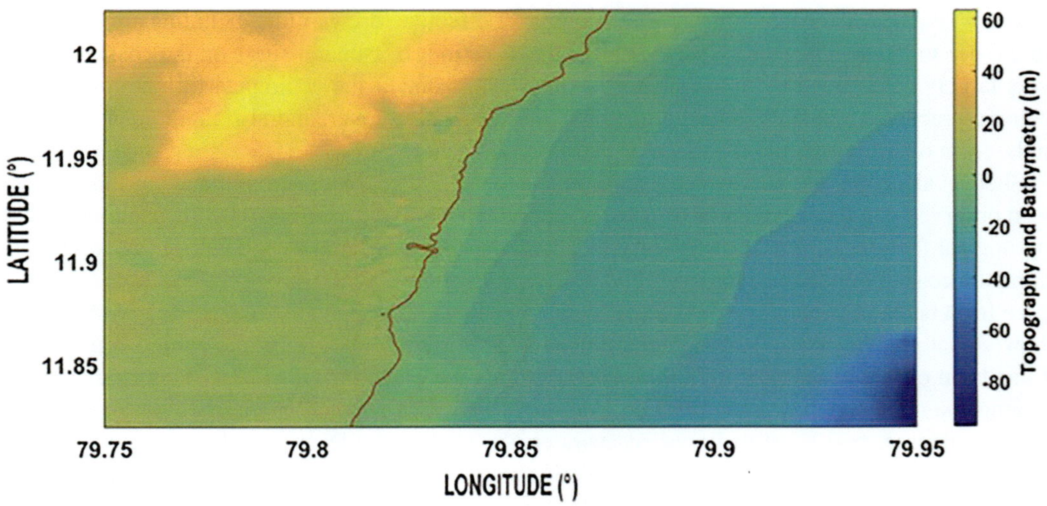

FIGURE 4.1 Map of Puducherry coast.

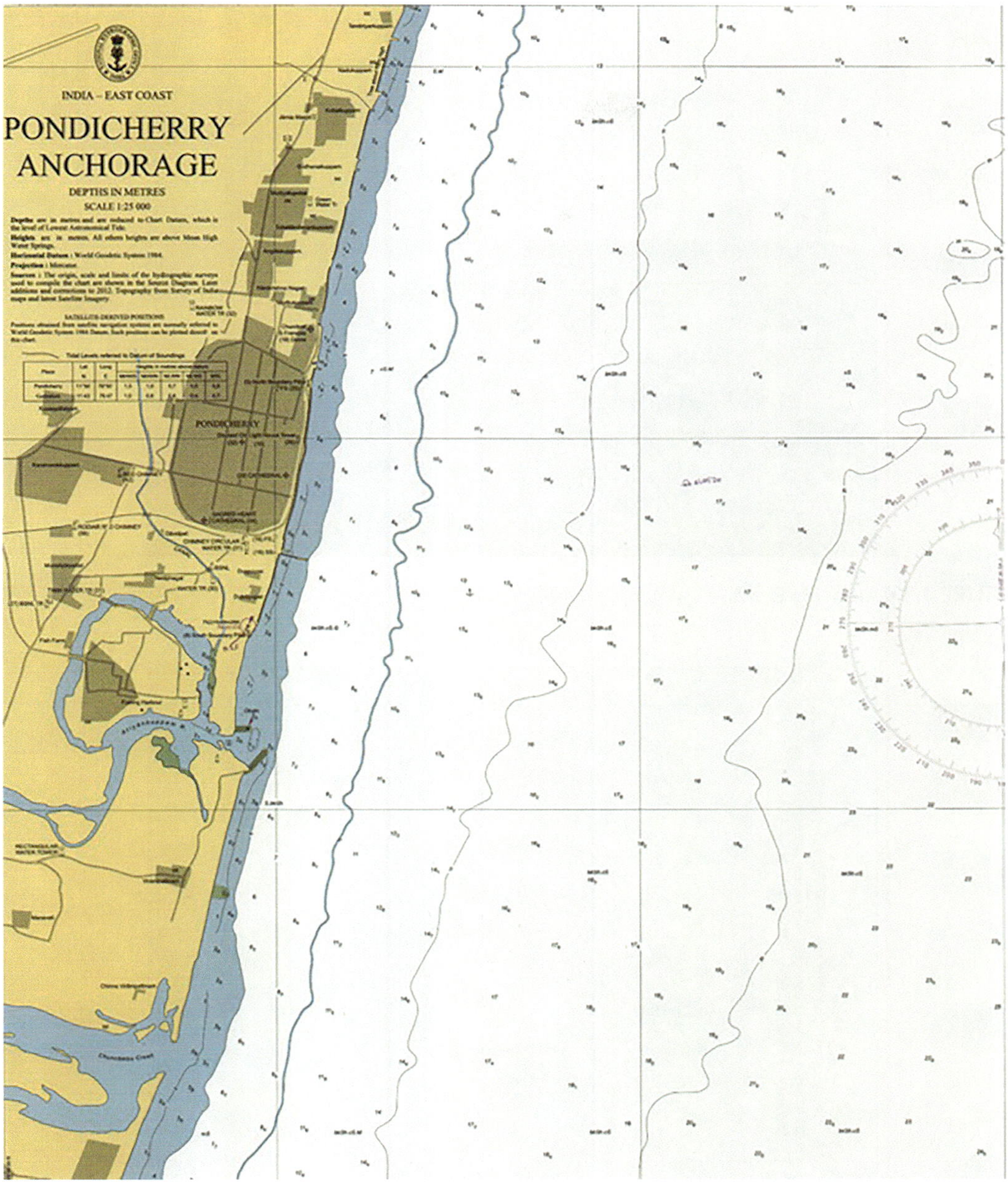

FIGURE 4.2 Topography and the Bathymetry of the coast.

The wave heights along Auroville Beach (2.5 m approx.) are much less compared to Paradise Beach (3.5 m approx.). Auroville Beach along the northern part of the coast has a very steep slope towards the sea (Figure 4.3). This clearly explains the decrease in inundation near the beach. Paradise Beach (Figure 4.5) is a relatively flat beach along the Puducherry coast. Not many structures are seen near the beach. This may be due to its gentle slope towards the sea, resulting in high inundation during tsunami or storm surges.

We can observe that along the three different beaches, the run-up and inundation vary.

FIGURE 4.3 Auroville Beach.

FIGURE 4.4 Beach Road.

FIGURE 4.5 Paradise Beach.

FIGURE 4.6 A road-sign with tsunami evacuation route in Puducherry.

- Beach Road: There is no inundation, but the run-up goes from 3 to 3.5 m. This can be explained using the presence of rock boulders along the coastline.
- Auroville Beach: Here inundation is observed, but the wave heights are low compared to the other two regions, ranging from 2.5 to 3 m.
- Paradise Beach: Both inundation and high run-ups (3.4–4 m) are observed here (Table 4.2).

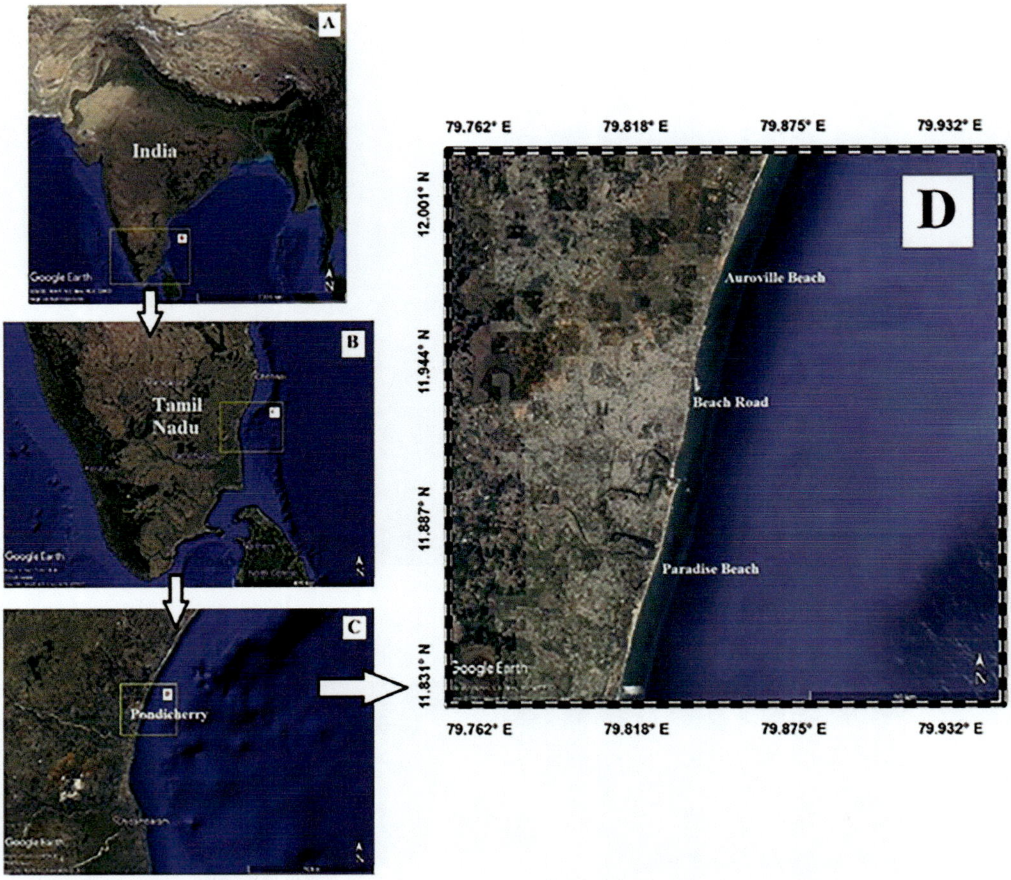

FIGURE 4.7 Location of study area.

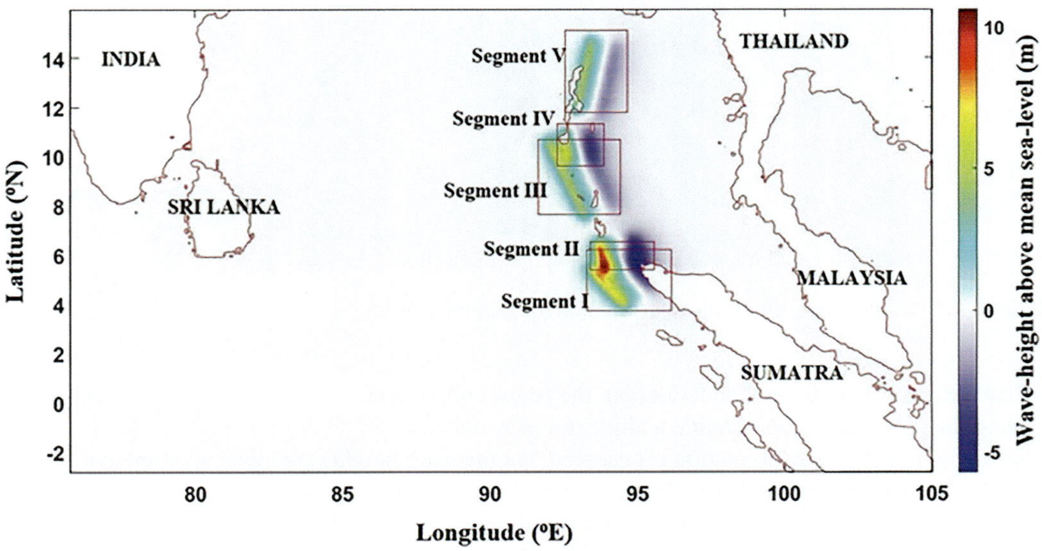

FIGURE 4.8 Deformation map for 2004 Indian Ocean tsunami.

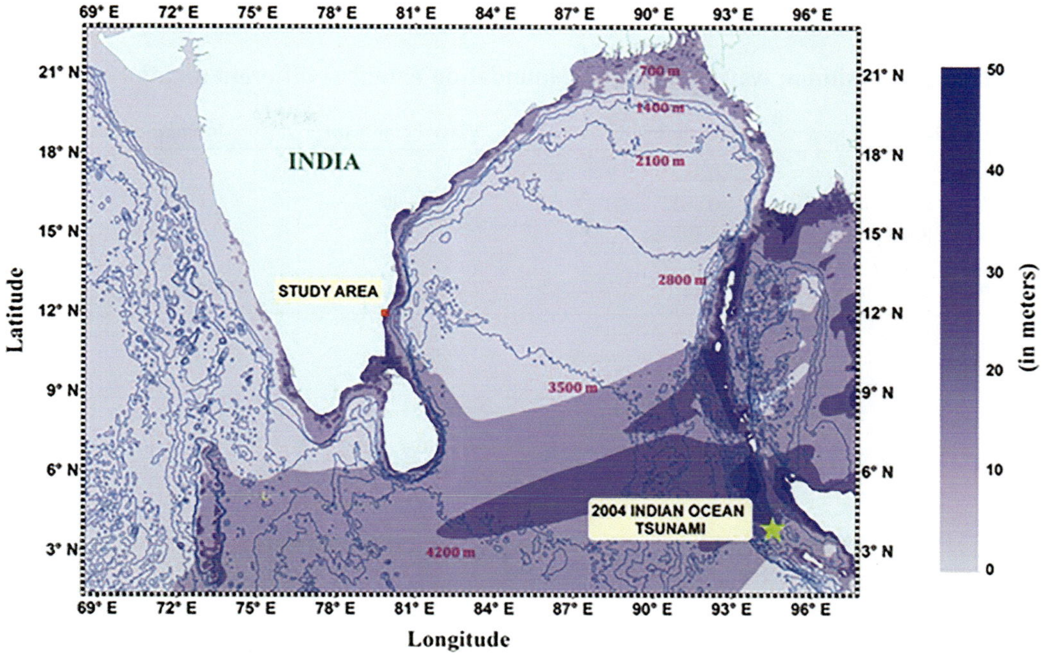

FIGURE 4.9 Directivity map for 2004 Indian Ocean tsunami.

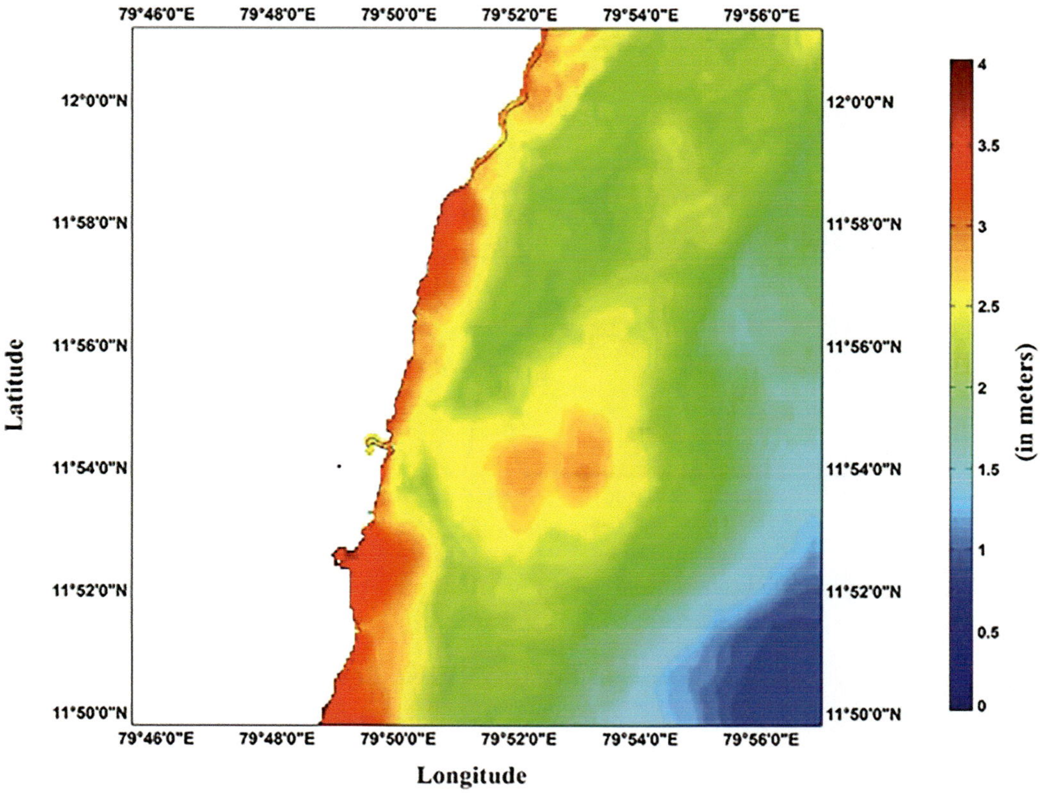

FIGURE 4.10 Run-up along Puducherry coast due to 2004 Indian Ocean tsunami.

TABLE 4.2
Depths, maximum wave heights, and inundation extent at different locations

Locations	Depth (m)	Wave heights (m)	Inundation (m)
Auroville Beach	-0.387	2.72	90
Beach Road 2	-0.3622	3.29	0
Beach Road 2	-2.79848	3.07	90
Beach Road 3	-2.32299	2.97	90
Paradise Beach	-2.84	3.62	630

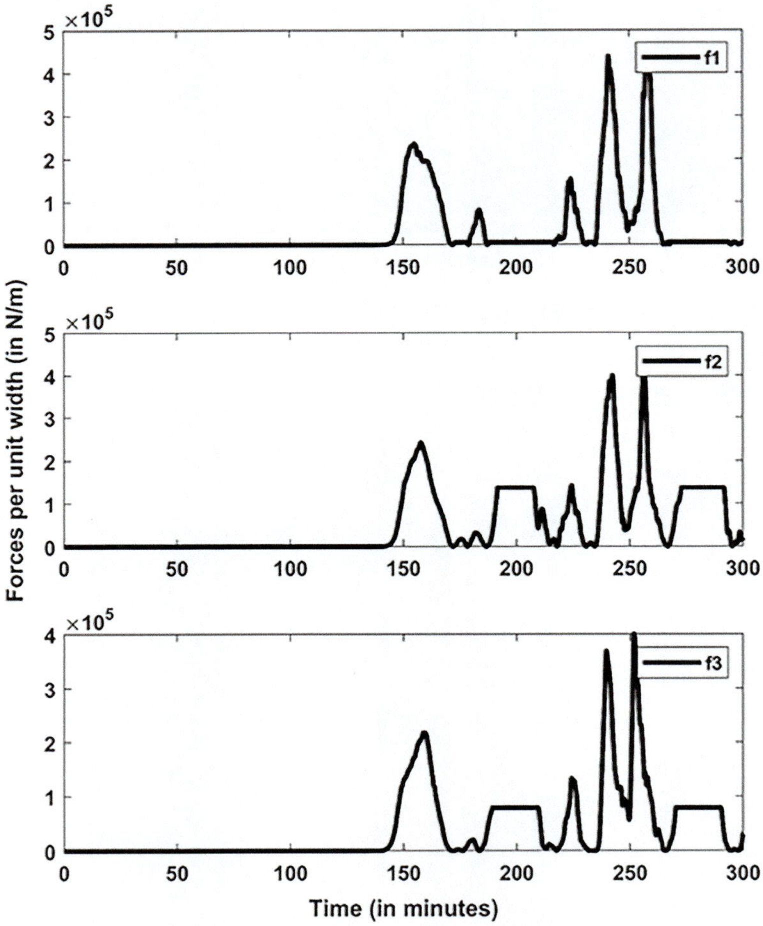

FIGURE 4.11 Force per unit width vs Time along Beach Road.

4.4.3 Tsunami-Induced Forces on the Boulders along Beach Road

Another study was conducted to assess the efficacy of rock boulders installed along the shoreline in reducing the impact of tsunamis. Beach Road (Figure 4.4), which is the most populated part of the Puducherry coast, has many buildings and infrastructure within a close distance to the sea and is thus extremely vulnerable to tsunami risks. Rock boulders installed along this section work like a seawall. As stated earlier, in our computation, such boulders are represented by a continuous seawall

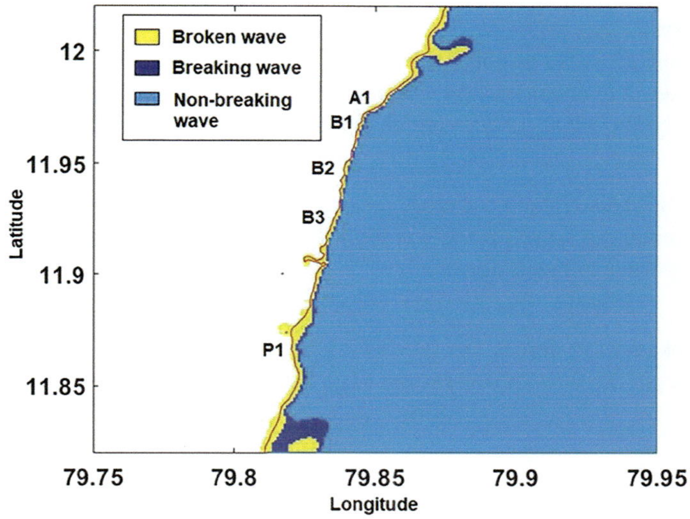

FIGURE 4.12 Categorization of Puducherry bathymetry based on 2004 Indian Ocean tsunami wave-heights.

to compute the impact forces. When tsunami waves move ashore, they apply a surge force on any hindrance to their path. Per unit width surge force acting on a building is computed employing the method outlined by Dames and Moore (1980) as

$$F / w = 4.5\rho g h^2$$

where ρ is the density of water, g is the gravitational acceleration, and h is the height of the wave.

Force per unit width is plotted with respect to time (Figure 4.11). It is observed that the force per unit width is of the order 20^5. The maximum impact is observed around four hours after the earthquake.

We understand that the force exerted on the sea wall is not similar to that on the boulders. Due to their structure, the waves hit the boulders in different directions causing the wave energy to dissipate in all directions. Eventually, the surge wave breaks and loses its velocity. Hence, no inundation or a slight overflow is observed. But, in the case of sea walls, the entire energy is transferred to the wall, thereby causing breakage in the walls or completely destroying the wall.

It is observed that Paradise Beach (Figure 4.5) has a relatively flat coast compared to the steep slope of Auroville Beach (Figure 4.3) which causes the run-up wave to go higher. There are rock boulders all along Beach Road (Figure 4.4). For all the three different beaches the tsunami wave heights and inundation extents are computed.

4.4.4 Vulnerability Map of Puducherry Coast

4.4.4.1 Categorization of Bathymetric Data based on Maximum Wave-heights

Based on the wave breaking criteria mentioned in Section 4.2.2, we divide the bathymetry of the Puducherry coast into three regions: Non-breaking, Breaking, and Broken wave regions Figure 4.12.

Using the formulas mentioned in Section 4.2, we have calculated the force/pressure for each region separately.

The maximum pressure calculated in the Non-breaking and Breaking wave regions are the pressures acting on the structures at the sea level.

Non-breaking waves:

Force due to non-breaking wave primarily consists of hydrostatic force due to the depth of the water. These forces are taken into consideration while building any deep-sea structure. In Figure 4.13, we can see the central region where the pressure is the highest.

Breaking waves:

The region south of Paradise Beach has more pressure than its northern parts due to its gentle slope (Figure 4.14). The breaking wave pressure is highest along Auroville Beach since the steep slope causes the incoming wave to break down.

Broken waves:

In Figure 4.15, broken wave forces are much higher along Paradise Beach than in the northern regions. These waves in the form of surges tend to go farther into the land with higher velocities due

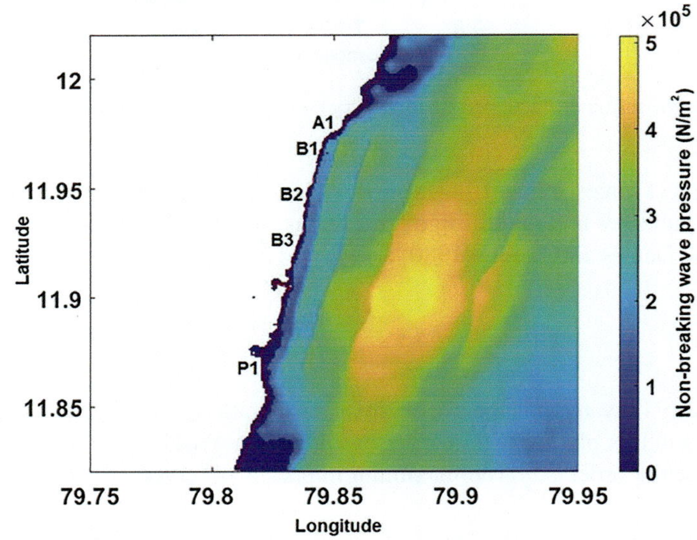

FIGURE 4.13 Plot of maximum non-breaking wave pressure along the non-breaking wave region.

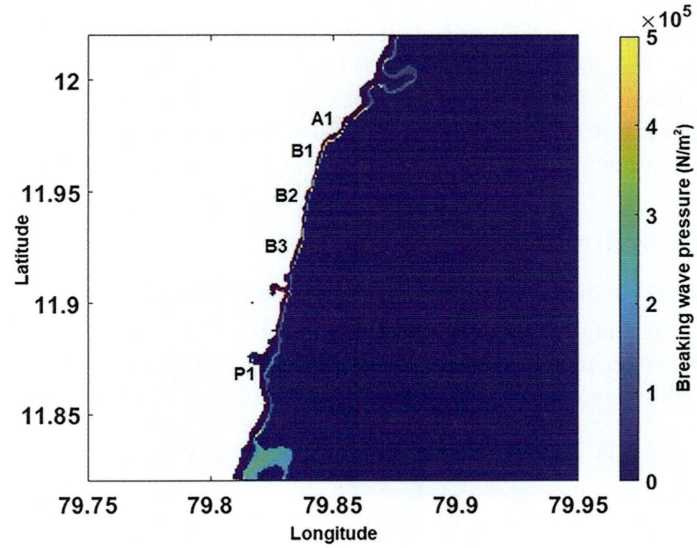

FIGURE 4.14 Plot of maximum breaking wave pressure along the breaking wave region.

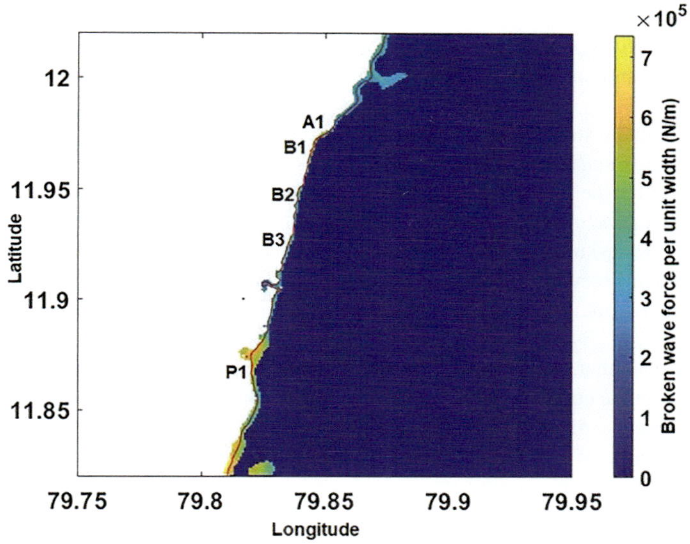

FIGURE 4.15 Plot of maximum broken wave forces along the broken wave region.

to the gentle slope. In the regions with a steep slope, like the Auroville Beach, the wave velocity is less and thereby the wave force is less.

4.5 SUMMARY

Tsunamis inflict catastrophic damage on coastal communities and economies. While early-warning systems and evacuation shelters have proven essential in mitigating loss of life, reducing the destruction of man-made structures remains the next critical challenge. The 2004 Indian Ocean tsunami devastated India's east coast—particularly the Nagapattinam district—and even well-prepared regions like Puducherry suffered extensive losses despite the existence of the National Tsunami Early Warning System. In a developing country such as India, safeguarding coastal infrastructure is vital to protecting both livelihoods and economic stability.

Deep sea structures generally withstand tsunami forces more effectively, since their design already accounts for hydrostatic and non-breaking wave pressures. Nevertheless, mapping the spatial distribution of maximum non-breaking wave pressure provides invaluable guidance for positioning future deep-sea structures. Closer to shore, however, breaking and broken wave forces dominate—and these can be far more severe. For example, Paradise Beach experiences significantly higher impact forces than neighboring stretches of coastline.

To comprehensively assess the vulnerability of Puducherry's coastal zone, it is therefore necessary to simulate a variety of tsunami scenarios. By combining pressure-distribution plots for both non-breaking and breaking waves with region-specific bathymetry and coastal morphology, engineers can identify the most resilient locations for critical infrastructure—and ultimately mitigate the economic and social fallout of future tsunami events.

REFERENCES

Anandabhaskaran. V. (2017). Monitoring shoreline changes of the Puducherry coast, South India: A review and a case study. *International Journal for Research in Applied Science and Engineering Technology*, *5*, 2470–2477. doi: 10.22214/ijraset.2017.11345

Azadbakht, M., & Yim S. C. (2014). Simulation and estimation of tsunami loads on bridge superstructures. *Journal of Waterway Port Coastal and Ocean Engineering*, *141*(2), 04014031. doi: 10.1061/(ASCE)WW.1943-5460.0000262

Chock, G., Robertson, I., Carden, L., & Yu, G. (2012). *Tohoku tsunami-induced building damage analysis including the contribution of earthquake resistant design to tsunami resilience of multi-story buildings.* Proceedings of the International Symposium on Engineering Lessons Learned from the 2011 Great East Japan Earthquake.

Cross, R. H. (1967). Tsunami surge forces. *Journal of Waterways and Harbor Division, 93*(4), 201–234.

Dames, M. (1980). *Design and construction standards for residential construction in tsunami-prone areas in Hawaii.* Prepared for the Federal Emergency Management Agency, Washington, DC.

Deo, M. C. (2013). *Waves and structures.* www.civil.iitb.ac.in/~mcdeo/waves.html

Elumalai, S., Victor, G., Rajeskumar, R., Soosai, D., & Chandrasekar, R. (2006). Damage assessment in the tsunami affected Puducherry – Cuddalore coast, India. In S. M. Ramasamy, C. J. Kumanan, R. Sivakumar, and Bhoop Singh (Eds.), *Geomatics in tsunami* (pp. 87–106). New India Publishing Agency, New Delhi, India.

Kumaraguru, A. K., Jayakumar, K., Wilson, J., & Ramakrishnan, C. M. (2005). Impact of the tsunami of 26 December 2004 on the coral reef environment of Gulf of Mannar and Palk Bay in the southeast coast of India. *Current Science, 89*(10), 1729–1741.

Kunkel, C. M., Hallberg, R. W., & Oppenheimer, M. (2006). Coral reefs reduce tsunami impact in model simulations. *Geophysical Research Letters, 33*, L23612. doi:10.1029/2006GL027892

Mizutani S., & Imamura, F. (2001). Dynamic wave force of tsunamis acting on a structure. *ITS 2001 Proceedings*, Session 7, No. 7-28.

Murali, M., Misra, A., Amrita, S., & Vethamony, P. (2013). Coastal vulnerability assessment of Puducherry coast, India, using the analytical hierarchical process. *Natural Hazards and Earth System Sciences. 13*, 509–559. doi: 10.5194/nhess-13-3291-2013

Murthy, M., Usha, T. and Pari, Y., & Tirumaleswara Reddy, N. (2011). Tsunami vulnerability assessment of Cuddalore using numerical model and GIS. *Marine Geodesy, 34*, 16–28. doi: 10.1080/01490419.2011.547797

Narayanan, R. M., Sharmila, K. J., & Dharanirajan, K. (2015). Bathymetry and Sea Floor Characteristics of Cuddalore and Puducherry Coast -India. *International Journal of Earth Sciences and Engineering, 8.* 364–371.

Nistor, I., Palermo, D., Nouri, Y., Murty, T., & Saatcioglu, M. (2009). Tsunami-induced forces on structures. In Y. C. Kim (Ed.), *Handbook of coastal and ocean engineering* (pp. 261–286). World Scientific, Singapore. doi: 10.1142/9789812819307_0011

Nouri, Y., Nistor, I., Palermo, D., & Cornett, A. (2010). Experimental investigation of tsunami impact on free standing structures. *Coastal Engineering Journal, 52*, 865–879. doi: 10.1142/S0578563410002117

Palermo, D., Nistor, I., Al-Faesly, T., & Cornett, A. (2012). Impact of tsunami forces on structures: The University of Ottawa experience. Proceedings of the Fifth International Tsunami Symposium (ISPRA-2012) Tsunami Society International, Ispra, 3–5 September.

Ramasamy, S. M. (2014). Imprints of Asian mega tsunami 2004 along Tamil Nadu – Puducherry coast, India and its significance. *International Geoinformatics Research and Development Journal, 5*, 21–28.

Robertson, I., Paczkowski, K., Riggs, H. R., & Mohamed, A. (2013). Experimental investigation of tsunami bore forces on vertical walls. *Journal of Offshore Mechanics and Arctic Engineering, 135*(2), 021621. doi: 10.1115/1.4023149.

Sainflou, M. (1928). Essai sur les digues maritimes verticales (Test on vertical sea dikes): Annales des Ponts et Chaussées, *98*, 5–48.

Sheth, A., Sanyal, S., Jaiswal, A., & Gandhi, P. (2006). Effects of the December 2004 Indian Ocean tsunami on the Indian mainland. *Earthquake Spectra, 22*, 435–473.

Suzuki T., Shijo, R., Yokoyama, K., Ikesue, S., Yamasaki, H., & Motoyama, J. (2014). Clarification of behavior of huge tsunami action on bridges – Hydraulic model experiment and simulation technology. *Mitsubishi Heavy Industries Technical Review, 51*(3), 44–49.

Swaroopa Rani, V., Srivastava, K., & Dimri, V. P. (2011). Tsunami propagation and inundation due to tsunamigenic earthquakes in the Sumatra-Andaman subduction zone: Impact at Visakhapatnam. *Marine Geodesy, 34*, 48–58. doi: 10.1080/01490419.2011.547802

5 Tsunami due to Meteorite Impact

5.1 INTRODUCTION

One cause of tsunamis is a meteorite impact in an ocean. It is rare for an asteroid or a meteorite to reach the Earth's surface and no tsunamis due to an asteroid impact have been documented in the past. However, in such an event, the effects would be disastrous. Although most meteorites burn out and disintegrate before reaching the Earth's surface, large meteorites have hit the Earth in the past (clearly evident from the large craters found in different parts of the world). One such example is the asteroid which may have fallen in prehistoric times—in the Cretaceous period, 65 million years ago—and this gigantic meteorite impact resulted in the creation of the Chicxulub crater. This is also called the K-T impact event (Cretaceous-Tertiary impact event). After discovering the Chicxulub crater on the Yucatan Peninsula, many studies of this Cretaceous-Tertiary event were related to its environmental impact. One of the studies was related to the tsunami deposits found in and around the Gulf of Mexico. Although their origin is still controversial, some discoveries have supported a tsunami generation by the Chicxulub impact event. Analysis of the seismic data gathered from the impact crater's offshore area indicates that the transient cavity's diameter is approximately 100 km (Morgan et al., 1997). According to Hildebrand et al. (1991), a circular structure with a diameter of 180 km located on the Yucatan Peninsula in Mexico, as an impact crater, is seen from gravity and magnetic studies. Matsui et al. (2002) have carried out tsunami modeling and have shown that the wave height of the tsunami generated by the crater is dependent on the depth of water at that time in the Yucatan Peninsula and also that the flow rate into the crater cavity is dependent on water depth of the Yucatan platform but not the depth of the crater. Since there is evidence of meteorite and asteroid impacts on our planet, we can deduce that there is a chance of an impact in oceans or seas, particularly since water covers four-fifths of the Earth's surface. During the past few decades, many scientists have estimated the hazards of meteorite impacts (Atkinson et al., 2000; Chapman and Morrison, 1994; Gehrels, 1994; Morrison, 1992), and it has been accepted that the possibility of a catastrophic experience caused by a rock from space is low.

5.2 METEORITE IMPACT PARAMETERS (1950 DA)

5.2.1 1950 DA and the March 16, 2880 Forecast

In 1950, the astronomers tracked an asteroid 1950 DA for 17 days and then lost it. Again in 2000, it was found headed inside the range of 20 lunar distances from Earth. The extrapolation of the orbit of 1950 DA by Giorgini et al. (2002) was more accurately done to date than any other near-Earth object. By merging the data from the highly accurate radar observations of the present velocity (±3.5 cm s^{-1}) and position (±0.9 km) with the original sightings of the asteroid, he concluded that on

DOI: 10.1201/9781003784500-5

March 16, 2880, the 1.1-km-diameter asteroid may impact Earth. However, before the impact event, the asteroid will have a few close confrontations with the Earth, Moon, and Mars. The calculated variance in each successive orbit widens due to the irregular perturbations caused by each encounter. From this Giorgini et al. (2002) concluded that the best-extrapolated position of 1950 DA in 2880 is variable to several lunar distances. He judged a possibility of an impact on Earth in 2880 to be less than 0.3 percent. Even though the impact likelihood of 1950 DA is low and has inaccuracies that may not be improved upon for centuries, Giorgini et al.'s (2002) citing of the date and time of a potential impact of this particular asteroid is widely accepted.

Giorgini et al. (2002) did not give any latitudes and longitudes for 1950 DA's point of impact with Earth but in this study, the coordinates near to Puducherry coast are considered.

5.3 MATHEMATICAL FORMULATION OF INITIAL WAVE AND PROPAGATION

5.3.1 Initial Wave and Propagation

In the early 2000s, there was a group looking at impact-generated waves using impact hydrocodes (iSALE). There was another group, however, that argued that linear theory is better for impact-generated tsunamis than the hydrocodes. That other group was Ward and Ashpaugh (Ward et al., 2000, 2003). The argument, in simple terms, is that hydrocode models are too dissipative over long (1000 km) distances. Matsui et al. (2002) used the iSALE hydrocode to calculate the initial wave generation and propagation of a tsunami from a Chicxulub impact event in the Gulf of Mexico. The differences between the two methods did not matter for Chicxulub in the relatively small confines of the Gulf of Mexico (unless one is also wanting to model waves moving over long distances and hitting Africa). Weiss (2006) presented a model, incorporating the iSALE impact model with a non-linear wave propagation model. Wünnemann et al. (2007) used a numerical model of hypervelocity impact to simulate several processes that accompany a comet or asteroid strike in the ocean or on the land. It stresses the wave generation mechanism and the initial phases of crater formation, such as the contact-compression phase, excavation phase, etc., are not given much importance. We use a similar approach in calculating initial wave generation.

Collision at such high speed is referred to as a hypervelocity impact because the impact velocity is many times faster than the speed of sound in water or rock. Hypervelocity impact results in the rapid compression of the impactor and target and the generation of a shock wave. A collision at such a high speed is referred to as a hypervelocity impact because the impact velocity is many times faster than the speed of sound in water or rock. Hypervelocity impact results in the rapid compression of the impactor and target and the generation of a shock wave.

5.3.2 Initial Wave Generation

Based on Wünnemann et al. (2010), a meteorite impact generates water waves through two different mechanisms. The first mechanism is rim uplift and collapse of the ejecta curtain and waves formed by this mechanism are called rim waves. The waves formed by the second mechanism, i.e., oscillations caused by cavity collapse, overshoot, collapse, overshoot, etc., are called collapse waves. Some studies suggest that rim waves only occur when the impactor diameter is more than half the water depth; however, other computations show the generation of a rim wave even in deep water impacts. In this study, we assume that during an oceanic impact, a rim wave is always created. But collapse waves are generated only in deep water impacts, i.e., when the impactor diameter is less than half the depth of the water layer.

Numerical impact simulations show that after an impact the transient crater is uplifted followed by the succeeding outward collapse of the curtain of material released from the crater finally forming a rim wave. It has been observed that the rim wave is soliton-like in form with a wavelength similar to transient crater diameter D_{tc} and reaches a maximum amplitude at a radial distance $R_{rw} = 3D_{tc}/4$. It is exactly at the center of the rim of the transient crater and the boundary of the zone inside which the majority of the ejecta lands. Collins et al. (2005) gives an appropriate estimation of the maximum rim wave amplitude,

$$A_{rw}^{max} = h_{tr} = 14.1 D_{tc} \tag{5.1}$$

Also, it is assumed that the amplitude of the wave cannot be more than the water depth. Hence when the impact occurs in shallow-water

$$A_{rw}^{max} = min\left(14.1\ D_{tc},\ H\right) \tag{5.2}$$

here H is the water depth.

Following the rim wave formation, we observe the collapse of the transient crater, later it overshoots the pre-impact surface, then a secondary crater is formed as it collapses, overshoots again, etc. These complex oscillations in water height near the impact zone generate the collapse waves of maximum amplitude $A_{cw}^{max} = 0.06 \min\left(d_{tc}, H\right)$, where d_{tc} is the transient crater depth in the

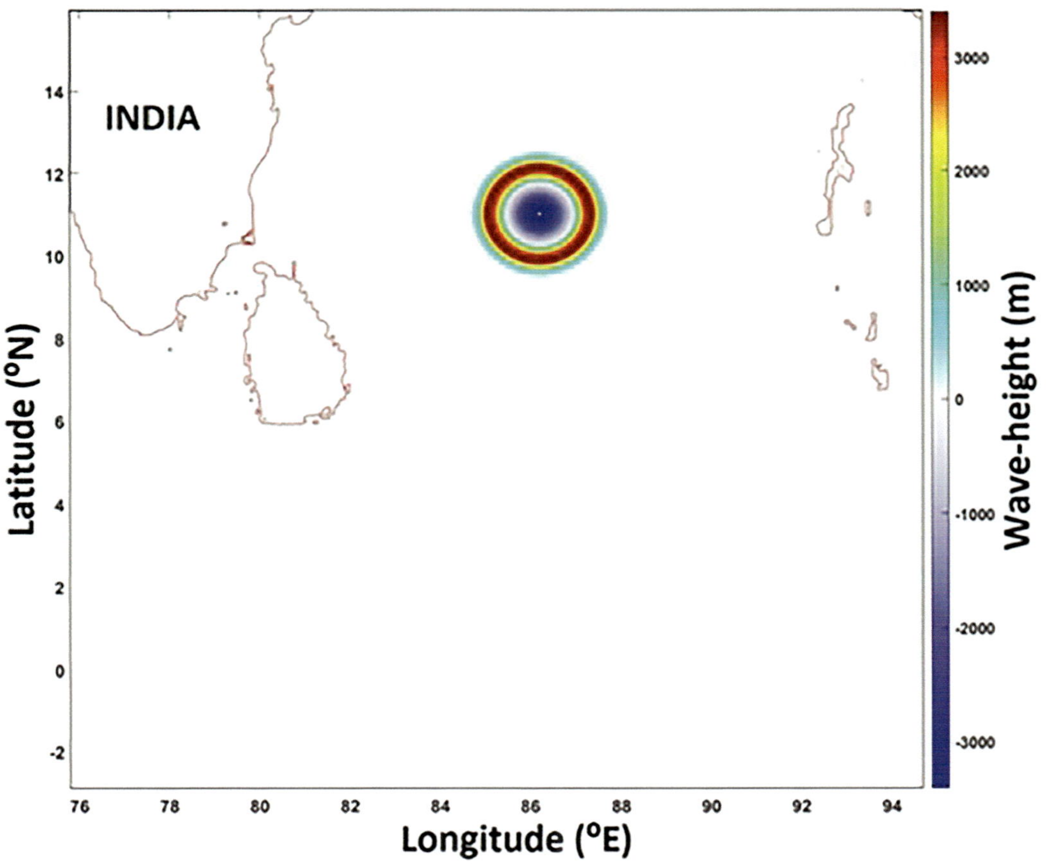

FIGURE 5.1 Initial wave due to meteorite impact.

water, $d_{tc} = D_{tc}/2.828$. It is assumed that the radial distance at which this maximum collapse-wave amplitude occurs is $R_{cw} = 5D_{tc}/2$.

5.3.3 Calculation of D_{TC}

It is not an easy task to determine the size of the transient crater from a particular impactor size, angle of incidence velocity, and density. Collins et al. (2005) has given the equation for calculating D_{tc}. He used a set of scaling laws that extrapolated the outcome of small-scale experimental data to required scales or to extend cratering survey on planets other than the Earth.

$$D_{tc} = 1.365\left(\frac{\rho_i}{\rho_t}\right)^{1/3} L^{0.78} v_i^{0.44} g_E^{-0.22} sin^{1/3}\theta \qquad (5.3)$$

where ρ_i is density of impactor, ρ_t is density of target(ocean), L is impactor diameter, v_i is impact velocity at the surface, g_E is acceleration due to gravity (9.8m/s^2), and θ is angle of impact.

TUNAMI N2 code is used for wave propagation with open sea and perfectly reflecting land as boundary conditions.

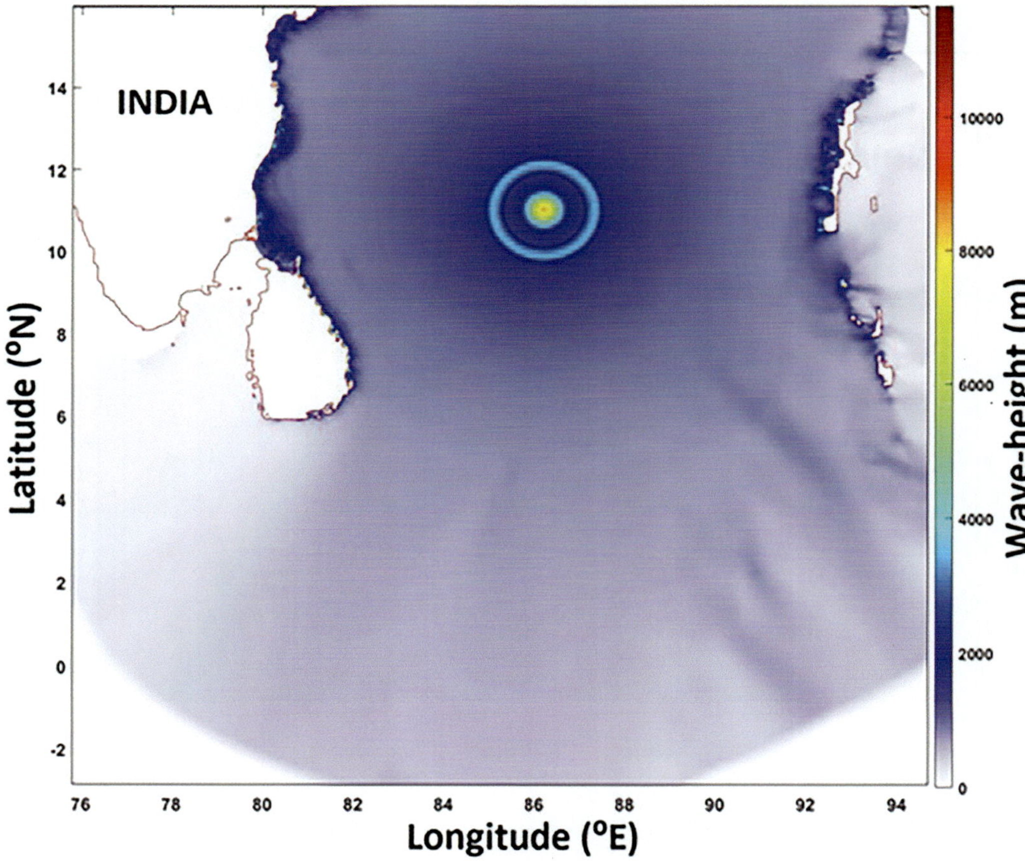

FIGURE 5.2 Directivity of meteorite impact tsunami.

5.4 RUN-UP AND INUNDATION ALONG THE COAST: EXAMPLE OF THE COASTAL REGION OF PUDUCHERRY, INDIA

The initial deformation (Figure 5.1) and directivity maps (Figure 5.2 for the input parameters as given in Table 5.1 were computed for a given meteorite impact in the Indian Ocean.
The three study regions of the Puducherry coast, India, were

(I) Auroville Beach
It is located in the northern region of Puducherry and it has a steeper slope which results in lower run-up heights and inundation.

TABLE 5.1
Impact parameters of 1950 DA with impact location in the Bay of Bengal

Coordinates	Depth of Ocean (m)	Density of water (kg/m³)	Density of impactor (kg/m³)	Diameter of impactor (m)	Velocity of impactor (m/s²)	Angle of impact (⁰)
86.17⁰E 10.99⁰N	3418	1.025	2200	1100	17800	45

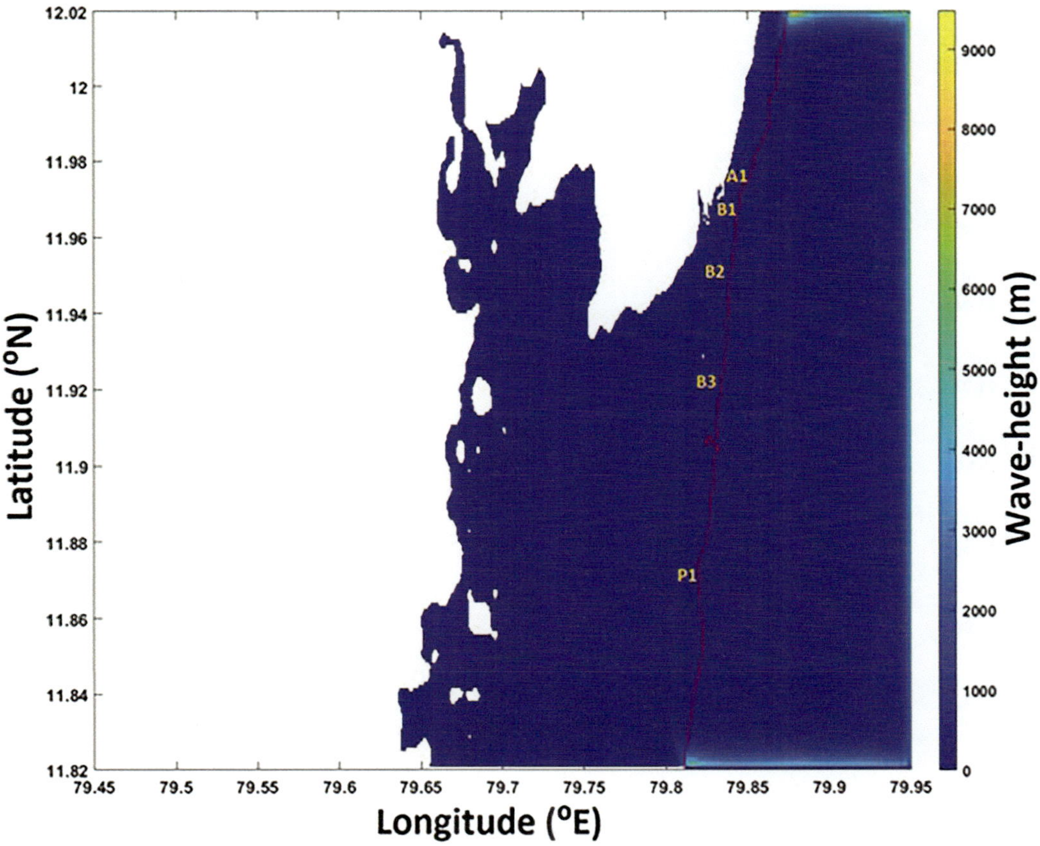

FIGURE 5.3 Run up along Puducherry Coast.

(II) Beach Road
Located in the central part of the Puducherry coast, this beach consists of rock boulders which act as a sea wall.

(III) Paradise Beach
Paradise beach is situated in the southern part of Puducherry coast with a relatively flat slope resulting in higher wave-heights and more inundation.

For all the three different beaches the tsunami wave heights and inundation extents are computed.

TABLE 5.2
Depths, maximum wave heights, and inundation extent at various locations

Locations	Depth(m)	Wave heights(m)	Inundation (m)
Auroville Beach, A1	-0.387	284	1530
Beach Road 1, B1	-0.3622	282.87	2430
Beach Road 2, B2	-1.79848	277.53	3780
Beach Road 3, B3	-1.31299	257.65	17100
Paradise Beach, P1	-1.84	233.18	15750

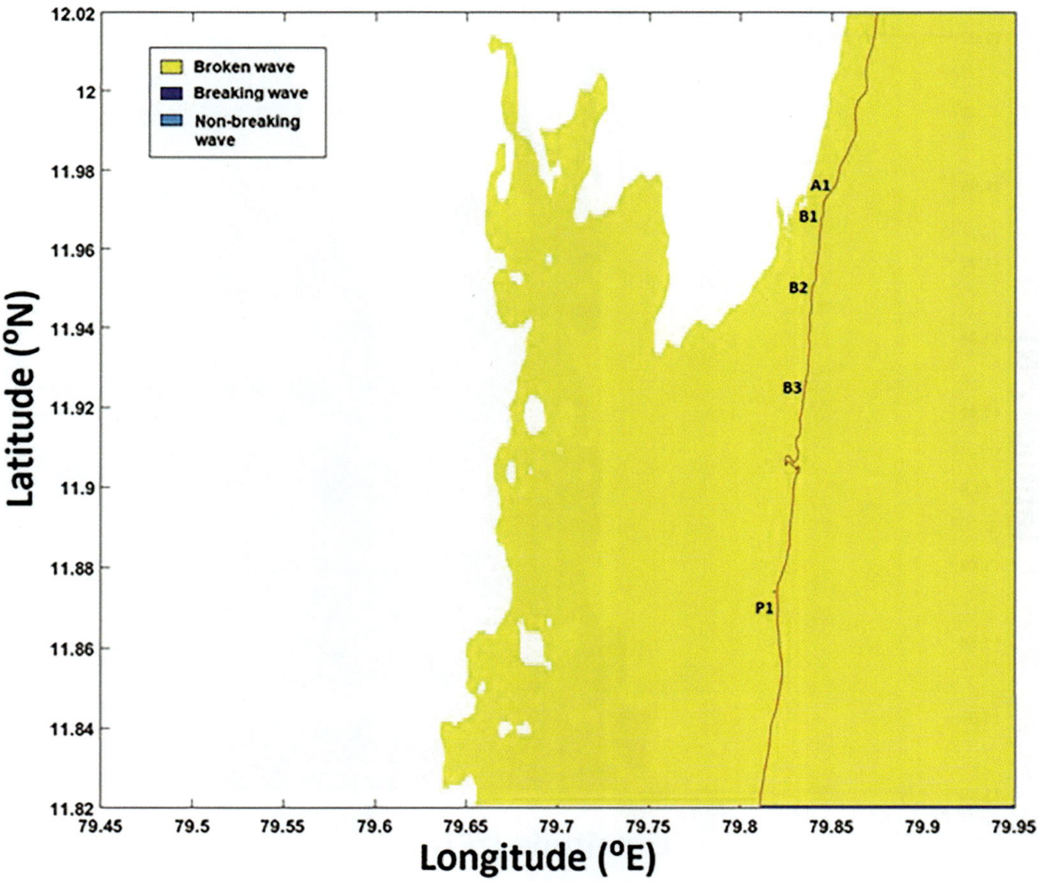

FIGURE 5.4 Categorization map of Puducherry Coast.

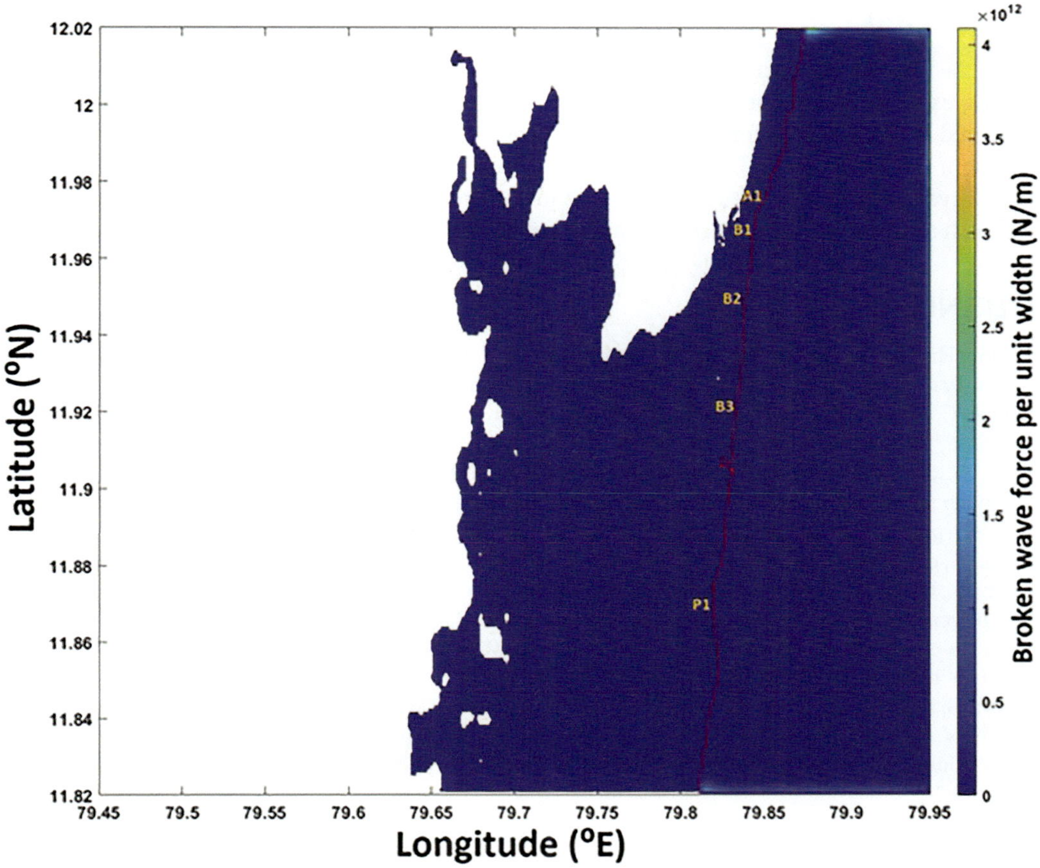

FIGURE 5.5 Broken wave force per unit width along Puducherry Coast.

TABLE 5.3
Broken wave forces per unit width at different locations

Locations	Broken wave force per unit width (10^9 N/m)
Auroville Beach, A1	3.653
Beach Road 1, B1	3.624
Beach Road 2, B2	3.4884
Beach Road 3, B3	3.0066
Paradise Beach, P1	2.4626

The run-up map of the Puducherry coast is given in Figure 5.3. The run-up wave heights and inundation extents at various locations are given in Table 5.2. From Figure 5.3 and Table 5.2, it is evident that the inundation along Auroville Beach and Beach Road is very much less compared to Paradise Beach. The inundation at B3 is high due to the waves from Paradise Beach.

5.4.1 Tsunami-Induced Forces

Figure 5.4 gives the categorization of Puducherry coast after the meteorite impact tsunami. It is observed that there are only broken waves in the study region. The broken wave forces are higher along the northern region Figure 5.5 and Table 5.3.

5.5 SUMMARY

Even if the meteorite impact-generated tsunami is a very rare event, the impact due to this is beyond anything expected. An event called the Chicxulub impact which occurred in Yucatan Peninsula is considered to be the reason for the extinction of many species, especially dinosaurs. Hence a deeper analysis in this direction will help in identifying locations for building safe houses for such catastrophic events. The occurrence of a meteorite impact-generated tsunami has a very low likelihood, but if such an event does ever occur it changes the face of any vulnerability studies done in the past.

REFERENCES

Atkinson, H., Tickell, C., & Williams, D. (2000). *Report of the task force on potentially hazardous near-earth objects*. Published by the United Kingdom NEO Information Centre, Cambridge.

Chapman C. R., & Morrison, D. (1994). Impact on Earth by asteroid and comets assessing the Hazard. *Nature*, *5*(367), 33–34. doi: 10.1038/367033a0

Collins, G. S., Melosh, H. J., & Marcus, R. A. (2005). Earth impact effects program: A Web-based computer program for calculating the regional environmental consequences of a meteoroid impact on Earth. *Meteoritics & Planetary Science*, *40*, 817–840.

Gehrels, T., ed. (1994). *Hazards due to comets and asteroids*. University of Arizona Press, Tucson, AZ.

Giorgini, J., Ostro, S., Benner, L., Chodas, P., Chesley, S., Hudson, S., Nolan, M., Klemola, A. R., Standish, E., Jurgens, R., Rose, R., Chamberlin, A., Yeomans, D., Margot, J. (2002). Asteroid 1950 DA's encounter with earth in 2880: Physical limits of collision probability prediction. *Science (New York, N.Y.), 296*, 132–136. doi: 10.1126/science.1068191

Hildebrand, A. R., Penfield, G. T., Kring, D. A., Pilkington, M., Camargo, Z. A., Jacobsen, S. B., & Boynton, W. V. (1991). Chicxulub crater: A possible Cretaceous/Tertiary boundary impact crater on the Yucata´n Peninsula, *Mexico: Geology, 19*, 867–871. doi: 10.1130/0091-7613(1991)019<0867:CCAPCT>2.3.CO;2

Matsui, T., Imamura, F., Tajika, E.,Nakano, Y., & Fujisawa, Y. (2002). Generation and propagation of a tsunami from the Cretaceous-Tertiary impact event. *Geological Society of America Special Paper Geological Society of America Special Paper*, *356*. 69–77. doi: 10.1130/0-8137-2356-6.69

Morgan, J. V., Warner, M., & the Chicxulub Working Group. (1997). Size and morphology of the Chicxulub impact crater. *Nature*, *390*, 472–476. doi: 10.1038/37291

Morrison D., (chair). (1992). *The spaceguard survey, report of NASA international near-earth-object detection workshop*, January 1992 prepared for NASA Office of Space Science and Technology.

Ward, S. N., & Asphaug, E. (2000). Asteroid impact tsunami: A probabilistic hazard assessment. *Icarus 145*, 64–78. doi: 10.1006/icar.2000.6336

Ward, S. N., & Asphaug, E. (2003). Asteroid impact tsunami of 2880 March 16. *Geophysical Journal International*, *153*(3), F6–F10. doi: 10.1046/j.1365-246X.2003.01944.x

Weiss, W., Wünnemann, K., & Bahlburg, H. (2006). Numerical modelling of generation, propagation and run-up of tsunamis caused by oceanic impacts: model strategy and technical solutions. *Geophysical Journal International*, *167*(1), 77–88. doi: 10.1111/j.1365-246X.2006.02889.x

Wünnemann, K., Weiss, R., & Hofmann, K. (2007). Characteristics of oceanic impact- induced large water waves – Re-evaluation of the tsunami hazard. *Meteoritics & Planetary Science*, *42*, 1893–1903. doi: 10.1111/j.1945-5100.2007.tb00548.x

Wünnemann, K., Collins, G. S., & Weiss, R. (2010). Impact of a cosmic body into earth's ocean and the generation of large tsunami waves: Insight from numerical modelling. *Reviews of Geophysics*, *48*(4), B06203, doi:10.1029/2009RG000308.

6 Tsunami due to Landslide

6.1 INTRODUCTION

A landslide tsunami is triggered by tectonic activity and instability caused by the ongoing sediment accumulation process. Moreover, if the volume of the landslide exceeds one cubic kilometer, it can cause a highly destructive tsunami. Maslin et al. (2004) and Chaytor et al. (2009) have shown that submarine landslides and sediment gravity flows can potentially displace substantial volumes of sediment beyond the continental slope. These mass movements can disturb and damage seafloor infrastructure (Parker et al., 2008; Yuan et al., 2012). Such a situation can cause a tsunami, resulting in a geo-hazard (Synolakis et al., 2002; Løvholt et al., 2006). Heidarzadeh et al. (2014) highlighted the unpredictable nature of landslide tsunamis in contrast to tectonic tsunamis, which are preceded by noticeable shaking associated with moderate to large earthquakes. Waves generated by landslides can occur suddenly and without any warning, e.g., the Skagway, Alaska, event in 1994, where the tsunami reached the coast within just one minute. As documented by Kulikov et al. (1996) and Synolakis et al. (2000), the rapid onset of the tsunami further emphasizes the urgency and limited time available for coastal communities to respond and evacuate immediately. A few extensively documented incidents have been crucial in drawing attention to landslide tsunamis. One such event occurred during the Flores Island earthquake in Indonesia in 1992, which Yeh et al. (1993) studied and found that the highest run-up was 26 m on Flores Island. Imamura and Imteaz (1995) studied the 1994 Mindoro strike-slip earthquake in the Philippines, which triggered a landslide tsunami which moved a 6,000-ton block about 1 km inland. It was only after the 1998 Papua New Guinea (PNG) catastrophic event that most of the other events were revisited, and the PNG tsunami was studied by several researchers (Synolakis et al., 2002). Some of the underwater landslide tsunamis triggered by earthquakes are the PNG tsunami in 1998, the Grand Banks tsunami of Canada in 1929 (Fine et al., 2005), and the Messina tsunami in Southern Italy in 1908 (Billi et al., 2008). Landslide tsunamis can be delayed by a few minutes to several tens of minutes after the main earthquake event. This delay can be attributed to the gradual failure of metastable sediment or to the impact of a minor secondary trigger, such as an aftershock, which upsets the delicate balance (Turner and Schuster, 1996).

A local landslide tsunami was generated on January 29, 2014 at Statland Namdalseld, mid Norway, with a run-up of 10 m, resulting in local damage (Glimsdal et al., 2016). Using the BING model given by Imran et al. (2001), different scenarios of landslides were run to compute the tsunami run-ups and compare the results with the observations. Micallef et al. (2016) pointed out more than 200 landslides in the Malta Escarpment, several of which are very small and less likely to generate a tsunami. The irregular continental shelf along the Maltese island is on the Malta Plateau's south-western edge. Micallef et al. (2016) studied a large landslide along the outer Malta Plateau, which is located upslope of Cumees Canyon, generating a 10 m wave height and reaching the coast in 18 minutes.

DOI: 10.1201/9781003784500-6

For the Indian coast, Jayraman (2009) has not ruled out the possibility of a landslide tsunami in both the Indus fan in the Arabian Sea and the Bengal fan in the Bay of Bengal. According to Rajendran et al. (2008), the northern Indian Ocean contains two essential components for submarine landslides: sediments and seismic faults. The Bengal fan sits on a vast reserve of gas hydrates and its dislocation could cause a landslide tsunami. Gas hydrates are ice-like substances made of water and natural gas, and their dissociation has been identified as a factor in other submarine slides (Beget et al., 2007). With this in mind, a study of a landslide tsunami offshore of the KG Basin is investigated for the hazard along the AP coast.

6.2 SOURCE MECHANISM

A solid block landslide model using a sliding function to describe the trajectory and speed of solid materials going down a slope is used. More details about the solid block model are given in Watts (1998), Watts et al. (2003), and Enet and Grilli (2007).

A submarine landslide causes a variation in the seafloor which has been extensively studied by Watts (1998), Watts et al. (2003), Liu et al. (2003), and Enet and Grilli (2007). In this model, it is assumed that a considerable volume of solid material shaped like an ellipsoid is displaced from its original position and slides down a slope, consequently changing the original bathymetry. Figure 6.1 illustrates the underwater slide of a solid rigid material coming down a straight inclined plane as shown by Watts et al. (2003). They have shown that the center of mass of the slide moves parallel to the incline and is subject to external forces from added mass, gravity, and dissipation. They assume a specific density of $\gamma \approx 1.85$, a negligible Coulomb friction coefficient of Cn ≈ 0, an added mass coefficient of Cm ≈ 1, and a drag coefficient of Cd ≈ 1; the motion of the underwater slide can be described using the following terms:

$$s(t) = s_0 ln\left[\cosh\left(\frac{t}{t_0}\right)\right] \tag{6.1}$$

$$a_0 \cong 0.30\ g\ sin\theta \tag{6.2}$$

$$u_t \cong 1.16\sqrt{b\ g\ sin\theta} \tag{6.3}$$

$$s_0 \equiv \frac{u_t^2}{a_0} \cong 4.48b \tag{6.4}$$

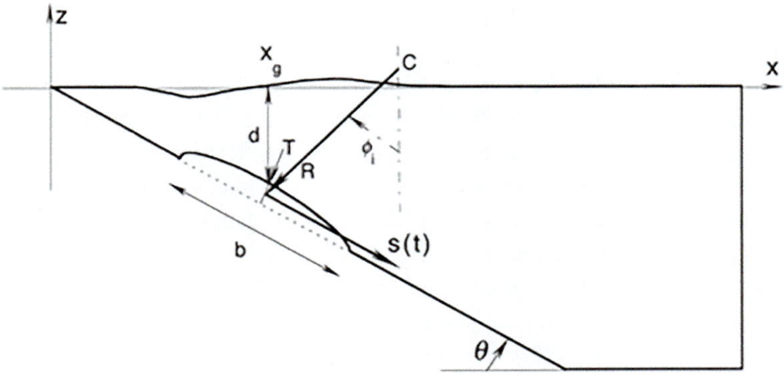

FIGURE 6.1 Schematic diagram of model for a underwater slide and slump.

$$t_0 \equiv \frac{u_t}{a_0} \cong 3.87\sqrt{\frac{b}{g\ sin\theta}} \quad (6.5)$$

where s(t) is center of mass motion, s_0 is characteristic distance of the motion, t_0 is characteristic time of the motion, a_0 is initial acceleration, and u is theoretical terminal velocity.

In contrast to instantaneous deformation where the duration of the rupturing process is negligible, transient seafloor motion involves a significant duration that cannot be ignored and the temporal variation needs to be modeled in a numerical simulation. The transient seafloor motion may be introduced by the slow rupture in an earthquake event or by submarine landslides. The insertion of temporal variation of bathymetry in governing equations allows simulation of tsunami waves induced by the transient floor motion process as long as the long wave assumption is valid. The model for transient seafloor motion has been successfully used to investigate the transient rupture process of the 2004 Indian Ocean tsunami, and the results showed a good match with satellite measurements (Wang and Liu, 2006).

6.3 CASE STUDY: POTENTIAL FOR SUBMARINE LANDSLIDES IN THE KG BASIN ON THE EAST COAST OF INDIA, NORTH INDIAN OCEAN

The KG continental basin was formed in the early Mesozoic due to the rifting of the eastern continental margin of India and is a petroliferous basin. The basin occupies about 28,000 km^2 on land, about 145,000 km^2 offshore, and lies between Visakhapatnam and Ongole from north to south. Rao (2001) has pointed out that the shelf and mid-slope of the offshore basin contain sediments ranging from the Carboniferous period to the Holocene epoch. On the other hand, the deeper sections of the basin, including the lower slope and areas further offshore, are predominantly filled with sediments from the Cretaceous to the Recent era (Curray et al., 1982; Pateria et al., 1992). After the Ganges and Brahmaputra, Godavari and Krishna are the largest rivers in Peninsular India, which transport the maximum sediment load to the Bay of Bengal; Godavari alone transports 50% of the total sediment load into the Bay of Bengal (BoB). According to Biksham and Subramanian (1988), this is the ninth-largest sediment transporting river in the world. The sediment load carried out by the Krishna and Godavari rivers ultimately gets deposited in the KG offshore basin located in the BoB along the eastern continental margin of India. Numerous researchers have devoted their efforts to investigating submarine mass transports and slope failures within the KG Basin. Based on extensive bathymetric and seismic profile surveys conducted by Ramprasad et al. (2011), several geological factors were identified, which include steep topographic gradients, high sedimentation rates, a regional fault system, diaprism, fluid/gas movement, and neo-tectonic activity. These conditions likely facilitated the slumping and sliding of sediments on the upper slope of the basin (Figure 6.1). The shale-dominated KG Basin, with proven gas hydrate deposits in these sedimentary settings, is mainly confined to continental slopes, and its dissociation might trigger submarine slope failures (Jyothsna et al., 2021). Krishna et al. (2016) reported that, based on analysis of long-streamer seismic reflection profile data, there was a significant shift in the direction of sediment flow within the basin.

Additionally, the study indicated a high sediment accumulation rate in the KG Basin exceeding 100 meters per million years during the time interval from 23 to 6.8 million years ago. A recent seismic fault activity was reported at 12:35:50 (IST) on August 24, 2021. This event resulted in an undersea earthquake with a magnitude of Mw 5.1 on the Richter scale, occurring at a sub-surface depth of approximately 10 km and the epicenter of this earthquake was located in the BoB at coordinates 14.440°N, 82.991°E. Using seismic reflection images of the BoB, Krishna et al. (2020) revealed that one of the rift-related faults within the KG Basin experienced reactivation during the Early Miocene period, approximately 16 million years ago. The fault recorded a cumulative throw of approximately 900 meters between the basement and the Early Miocene horizon (around 16 Ma).

Subsequently, the fault underwent reactivation at 6.8 Ma and has continued its activity progressively until approximately 0.3 Ma ago. The seismic activity and sedimentation rates make the region vulnerable to an underwater landslide, causing a tsunami. Several researchers have studied this basin extensively for sediment transport and slope failures. In the KG Basin, we consider a possible landslide model for different sediment flow structures and compute the run-up heights along the coast. To model the tsunami generated by landslides, the COMCOT numerical model is utilized, incorporating a realistic landslide scenario from the identified sliding zones to simulate the generation and propagation of tsunami waves, thereby computing the run-up height and inundation distances.

Based on the sliding and slumping phenomena observed in the KG Basin, Ramprasad et al. (2011) identified four distinct landslide sources, SLS-1, SLS-2, SLS-3, and SLS-4. Farveen et al. (2025) computed landslide tsunami studies for these four slides. However, in this study, we consider only two major possible landslide bodies from their research (Figure 6.2) to demonstrate the effect of landslide tsunami on the coast. The slope failure scars associated with these landslides is located in the north-western region of the KG Basin. The landslip scars are crescent-shaped, ranging in length from 4.85 to 10.6 kilometers and width from 4.7 to 7.2 kilometers. The sliding material volume within these scars ranges from 1.2 to 7.51 km^3. It is important to note that because there is no debris within the landslip scars, all the material is evacuated in a single event. The parameters used for modeling these events is provided in Table 6.1 where the slide volume is calculated for a landslides model as an ellipsoid block on a sliding surface.

6.4 SIMULATION OF LANDSLIDE TSUNAMI AND DISCUSSION

Tsunami simulations for the two different landslide sources, namely SLS-1, SLS-2, along with the corresponding data, are presented in Table 6.2. Figure 6.3 illustrates the distribution of maximum wave amplitudes both along the coast and in the ocean for the two scenarios and reveals that the highest tsunami waves are observed in the case of landslide SLS-2.

6.4.1 Analysis of Land Slide SLS-1

It is evident from Figure 6.3 that when landslide SLS-1 is modeled, we observe that at the source, the upliftment of the water column is around 7 meters. The primary direction of the wave movement is into the open sea and is towards the southern part of Andhra Pradesh (AP), while comparatively less energy is directed towards the northern areas of the AP coast. The maximum run-up height is observed at Antarvedi near the source region and is around 4.3 m. The waves reached different locations along the coastline within half an hour to two hours. The initial wave reached Antarvedi in just 30 minutes, followed by the arrival at Gangapatnam 4 minutes later with a wave height of 0.5 m. In 40 to 50 minutes, the wave reached Machilipatnam and Kottapatnam with wave heights of 2 meters and 0.4 meters, respectively. The wave arrival time and the run-up heights at different locations along the study region are given in Table 6.2. The waves reached as far as the northern parts of Tamil Nadu state, down south, with a maximum wave height of 2 meters.

6.4.2 Analysis of Land Slide SLS-2

The highest calculated run-up height of approximately 6.5 meters was generated by the landslide source SLS-2. From Figure 6.3, the initial displacement of the water column is about 10 meters, with wave energy distributed both along the coast and towards the open sea. In this landslide scenario, we observe wave propagation towards the north of the Tamil Nadu coast with run-up heights ranging from 1 to 2 m. Across the study region, the run-up heights varied from 0.3 to 6.5 m, and the waves reached different locations between 28 and 78 minutes. The highest wave height of 6.47 m was at Antarvedi after 28 minutes. It arrived at Gangapatnam with less than a 1 m wave height after

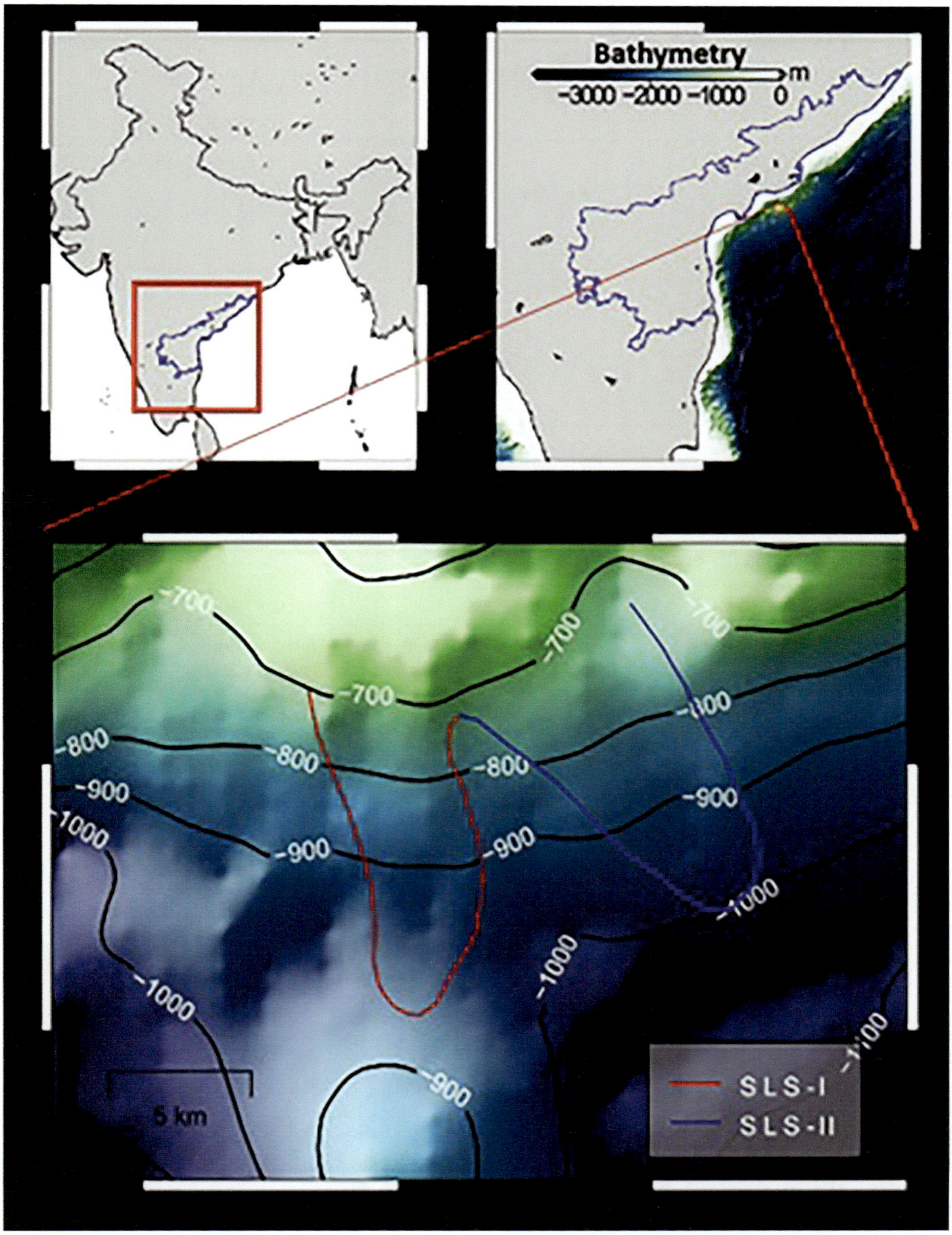

FIGURE 6.2 Map showing the regions of different flow structures marked in red and blue.

34 minutes. The plot illustrates the oscillations of the tsunami waves at selected points along the AP coast. The figure shows that tsunami waves comprise multiple trains, and at all locations, the initial wave is not the largest. Given that the source is located in the open ocean, the maximum wave energy dissipates towards the open ocean.

TABLE 6.1
Submarine landslide source (SLS) parameters considered in KG Basin

	SLS-1	SLS-2
Longitude (°) (start to stop)	81.8285 –81.8383	81.8815 – 81.945
Latitude (°) (start to stop)	16.01 – 15.918	16.0275 – 15.952
Average slope (°)	3.0	3.0
Bulk density (g/cm^3)	1.85	1.85
Length (km)	10.6	10
Width (km)	5.2	7.2
Thickness (km)	0.2	0.2
Volume (km^3)	5.75	7.51

TABLE 6.2
Maximum wave amplitudes and arrival times for SLS-1 and SLS-2 at selected numerical gauge points along with coordinates

			SLS-1		SLS-2	
Location	Lat (°E)	Long. (°N)	Arrival Time of initial wave (min)	Max. wave amplitude (m)	Arrival Time of initial wave (min)	Max. wave amplitude (m)
Visakhapatnam	17.70	83.35	72	0.60	50.8	1.63
Uppada	17.05	82.37	116.75	0.23	74.30	0.60
Antarvedi	16.28	81.74	29.35	4.25	28.4	6.47
Machilipatnam	16.15	81.24	48.38	1.66	46.97	4.95
Suryalanka	15.79	80.55	76.6	0.57	76.95	0.40
Kottapatnam	15.39	80.18	45.2	0.36	78.45	0.32
Gangapatnam	14.52	80.20	34	0.53	34.66	0.92

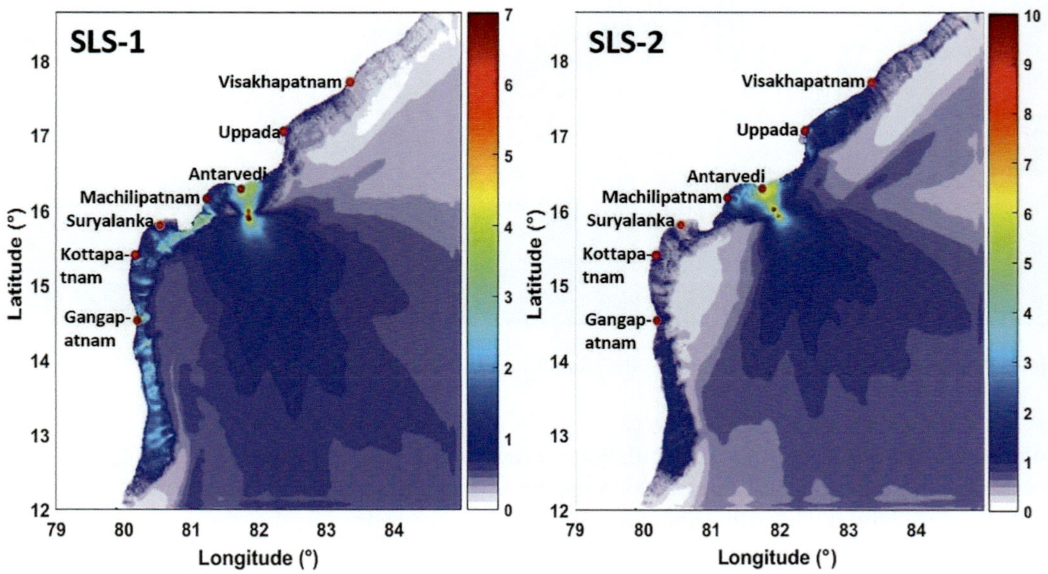

FIGURE 6.3 Distribution of maximum wave amplitudes in SLS-1 and SLS-2 along the coast and in the ocean.

6.5 SUMMARY

Landslide tsunamis can manifest in a unique environment with an elevated sedimentation rate. A considerable wave height in a landslide tsunami can cause significant damage and lead to loss of life. A case of a hypothetical submarine landslide in the offshore KG Basin is modeled for a landslide tsunami. Seismic section and bathymetry have revealed evidence of slumping and sliding in the region. A tsunami model for such a possible event is executed using the COMCOT model, and the tsunami arrival times and wave heights along the AP coast are quantified. At Antarvedi, the run-ups range between 4.5 and 6.5 m, and at Machilipatnam, they range between 1.6 and 5 m. The region's coastal areas are densely populated, and the presence of such large run-up heights can pose a significant hazard. The maximum run-ups with faster arrivals are observed in landslide SLS-2 scenarios, which could be because of the volume of the material, i.e. 7.51 km^3, and in the case of landslide SLS-1, the volume of material is less and is around 5.75 km^3. The effect of these slope failure-generated waves is confined to nearby source locations. A significant run-up height has the potential to cause major damage, depending on the topography of the region, and can also result in fatalities.

REFERENCES

Beget, J. E. (2007). Methane gas release from the Storegga submarine landslide linked to early Holocene climate change. *Holocene, 17*, 291–295.

Begum, F., Srivastava, K., Sahoo, R., Rajaraman, R., & Krishna, K.S. (2025). Numerical modeling of tsunami waves due to possible submarine landslides in Krishna-Godavari basin, Bay of Bengal. *Journal of Earthquakes and Tsunamis*, *117*, 187–196.

Biksham, G., & Subramanian, V. (1988). Sediment transport of the Godavari river basin and its controlling factors. *Journal of Hydrology*, *101*(1–4), 275–290. doi: 10.1016/0022-1694(88)90040-6

Billi, A., Funiciello, R., Minelli, L., Faccenna, C., Neri, G., Orechhio, B., and Presti, D. (2008). Landslide triggered Italian tsunami previously linked to quake. *Geophysical Research Letters*, *35*, L06301. doi:10.1029/2008GL033251

Chaytor, J. D., Ten Brink, U. S., Solow, A. R., & Andrews, B. D. (2009). Size distribution of submarine landslides along the U.S. Atlantic margin and its implication to tsunami hazards. *Marine Geology*, *264*, 16–27. doi: 10.1016/j.margeo.2008.08.007

Curray, J. R., Emmel, F. J., Moore, D. G., and Raitt, R. W. 1982. Structure, tectonics and geological history of the northeastern Indian Ocean. In A. E. Nairn & F. G. Stehli (Eds.), *The ocean basins and margins* (Vol. 6, pp. 399–450). Plenum Press, New York.

Enet, F., & Grilli, S. T. (2007). Experimental study of tsunami generation by three dimensional rigid underwater landslides. *Journal of Waterway, Port, Coastal, and Ocean Engineering*, *133*(6), 442–454. doi: 10.1061/(ASCE)0733-950X(2007)133:6(442)

Fine, I. V., Rabinovich, A. B., Bornhold, B. D., Thomson, R. E., & Kulikov, E. A. (2005). The Grand Banks landslide-generated tsunami of November 18, 1929: preliminary analysis and numerical modeling. *Marine Geology*, *215*(1–2), 45–57.

Glimsdal, S., L'Heureux, J. S., Harbitz, C. B., & Løvholt, F. (2016). The 29th January 2014 submarine landslide at Statland, Norway—Landslide dynamics, tsunami generation, and run-up. *Landslides*, *13*(6), 1435–1444.

Heidarzadeh, M., Krastel S., & Yalciner, A. C. (2014). The state-of-the-art numerical tools for modeling landslide Tsunamis: A short review. In S. Krastel et al. (Eds.), Submarine mass movements and their consequences, Chap. 43. *Advances in natural and technological hazards research* (Vol. 37, pp. 483–495). Springer, Cham.

Imamura, F., & Imteaz, M. M. A. (1995). Long waves in two-layers: Governing equations and numerical model. *Science of Tsunami Hazards*, *13*, 3–24.

Imran, J., Parker, G., Locat, J., & Lee, H. (2001). 1D numerical model of muddy subaqueous and subaerial debris flows. *Journal of Hydraulic Engineering*, *127*(11), 959–968.

Jayaraman, K. S. (2009). *Tsunami danger closer home*. Nature India. www.nature.com/articles/nindia.2009.8

Krishna, K. S., Ismaiel, M., Srinivas, K., & Saha, D. (2020). Post-breakup deformations in the Bay of Bengal: Response of crustal strata to the sediment load. *Journal of Earth System Science*, *129*, 159. doi: 10.1007/s12040-020-01436-7

Krishna, K. S., Ismaiel, M., Srinivas, K., Gopala Rao, D., Mishra, J., & Saha, D. (2016). Sediment pathways and emergence of Himalayan source material in the Bay of Bengal. *Current Science, 110*(3), 363–372. doi: 10.18520/cs/v110/i3/363-372

Kulikov, E. A., Rabinovich, A. B., Thomson, R. E., & Bornhold, B. D. (1996). The landslide tsunami of 3 November 1994, Skagway harbor, Alaska. *Journal of Geophysical Research, 101*(C3), 6609–6615.

Liu, P. L. F., Lynett, P., & Synolakis, C. E. (2003). Analytical solutions for forced long waves on a sloping beach. *Journal of Fluid Mechanics, 478*, 101–109.

Løvholt, F., Bungum, H., Harbitz, C.B., Glimsdal, S., Lindholm, C.D., & Pedeersen, G. (2006). Earthquake related tsunami hazard along the western coast of Thailand. *Natural Hazards and Earth System Sciences, 6*, 979–997.

Maslin, M., Owen, M., Day, S., & Long, D. (2004). Linking continental-slope failures and climate change: Testing the clathrate gun hypothesis. *Geology, 32*, 53–56. doi: 10.1130/G20114.1

Micallef, A., Georgiopoulou, A., Mountjoy, J., Huvenne, V. A., Iacono, C. L., Le Bas, T., ... & Otero, D. C. (2016). Outer shelf seafloor geomorphology along a carbonate escarpment: The eastern Malta Plateau, Mediterranean Sea. *Continental Shelf Research, 131*, 12–27.

Palle, J., Nittala, S., & Samudrala, K. (2021). Gas hydrate/free gas migration pathways in submarine slope failures: East Indian Margin. *Journal of Earth System Science, 130*, 83. doi:10.1007/s12040-021-01575-5.

Parker, E. K., Traverso, C., Moore, R., Evans, T., & Usher, N. (2008). Evaluation of landslide impact on deepwater submarine pipelines, Offshore Technology. *Conference,* 5–8 May, Houston, TX.

Pateria, M. L., Rangaraju, M. K., & Raiverman, V. (1992). A note on the structure and stratigraphy of Bay of Bengal sediments. In: Recent Geoscientific Studies in the Bay of Bengal and the Andaman Sea. *Geological Survey of India Special Publications*, *29*, 21–23.

Rajendran, C. P., Ramanamurthy, M. V., Reddy, N. T., & Rajendran, K. (2008). Hazard implications of the late arrival of the 1945 Makran tsunami. *Current Science*, *95*(12), 00113891.

Ramprasad, T., Dewangan, P., Ramana, M. V., Mazumdar, A., Karisiddaiah, S. M., Ramya, E. R., & Sriram, G. (2011). Evidence of slumping/sliding in Krishna Godavari offshore basin due to gas/fluid movements. *Marine and Petroleum Geology*, *28*, 1806–1816.

Rao, G. N. (2001). Sedimentation, stratigraphy, and petroleum potential of Krishna–Godavari basin, east coast of India. *American Association of Petroleum Geologists Bulletin, 85*(9), 1623–1643.

Synolakis, C. E., Bardet, J. P., Borrero, J. C., Davies, H. L., Okal, E. A., Silver, E. A., Sweet, S., & Tappin, D. R. (2002). The slump origin of the 1998 Papua New Guinea tsunami. *Proceedings of the Royal Society of London. Series A: Mathematical, Physical and Engineering Sciences*, *458*, 763–789.

Synolakis, C. E., Borrero, J. C., Plafker, G., Yalciner, A., Greene, G., & Watts, P. (2000). Modeling the 1994 Skagway, Alaska Tsunami. *Eos, Transactions American Geophysical Union*, *81*, F748 (Abstract).

Turner, A. K., & Schuster, R. L. (1996). Landslides: Investigation and mitigation, Special Report 247, Trans. Res. Board, National Academy Press, Washington, DC.

Wang, X. (9)). User manual for Cornell multi-grid coupled tsunami model-COMCOT V1.7. http://ceeserver.cee.cornell.edu/pll40/group/doc/COMCOT_User_Manual_v1_7.pdf (accessed 12 January 2017).

Wang, X., & Liu, P.L.F. (2006). An analysis of 2004 Sumatra earthquake fault plane mechanisms and Indian Ocean tsunami. *Journal of Hydraulic Research, 44*(2), 147–154.

Watts, P. (1998). Wavemaker curves for tsunamis generated by underwater landslides. *ASCE Journal of Waterway, Port, Coastal, and Ocean Engineering, 124*, 127–137.

Watts, P., Grilli, S. T., Kirby, J. T., Fryer, G. J., & Tappin, D. R. (2003). Landslide tsunami case studies using a Boussinesq model and a fully nonlinear tsunami generation model. *National Hazard Earth System, 3*, 391–402

Yeh, H., Imamura, F., Synolakis, C. E., Tsuji, Y., Liu, P., & Shi, S. (1993). The Flores Island tsunamis. *Eos, Transactions American Geophysical Union*, *74*(33), 371–373.

Yuan, F., Wang, L. Z., Guo, Z., & Shi, R. W. (2012). A refined analytical model for landslide or debris flow impact on pipelines. *Part – I Surface pipelines. Applied Ocean Research, 35*, 95–104.

7 Conclusions

7.1 GENERAL FINDINGS

This chapter discusses the significant findings of the research work and the scope for future work. This chapter presents a point-wise summary of all the chapters in the book.

Chapter 1 discusses the basics of tsunamis and their characteristics, and provides an overview of worldwide historical tsunamis. 85% of global tsunamis are due to underwater earthquakes, 8% are due to landslides, about 5% are due to volcanic activity, and fewer than 2% are due to meteorite impacts. These four causes are examined thoroughly, and the associated tsunamis are tabulated. As major global tsunamis are due to tsunamigenic earthquakes in subduction zones, special emphasis is given on the two main subduction zones in the Indian Ocean: the Makran subduction zone on the west of the Indian sub-continent and the Andaman-Sumatra subduction zone on the east. Descriptions of tsunami wave generation and wave propagation equations are presented in Chapter 2. This chapter discusses the initial tsunami wave formulation, propagation, run-up, and inundation for an underwater tsunamigenic earthquake. The TUNAMI-N2 code, which several researchers worldwide use, is presented. The accuracy of tsunami modeling and vulnerability mapping inherently depends on the quality and resolution of input data. Although high-resolution bathymetric and topographic data are being used, there are limitations due to the availability and accuracy of these datasets, especially in remote or less-studied areas. Variations in data quality can impact model precision and the reliability of predictions. Tsunami models rely on several assumptions and simplifications, such as idealized wave propagation and uniform bathymetric features.. The agreement between vulnerability estimates from different studies reinforces the reliability of such models for accurate risk assessments. Comparative studies in other tsunami-prone regions, such as Japan and Indonesia, have highlighted the importance of high-resolution data and dynamic modeling techniques.

In Chapter 3, different bathymetry features that have affected the tsunami propagation in the northern Indian Ocean are discussed. Studies highlighted that the 2004 Sumatra tsunami and the 1945 Makran tsunami necessitate prompt and efficient early warning systems to mitigate the impact on vulnerable coastal communities. The results of the earlier studies highlight the essential role of accurate tsunami modeling in improving response strategies and emphasizing the importance of natural barriers in mitigating tsunami impacts. Efforts to reduce tsunami vulnerability should focus on enhancing early warning systems, improving evacuation routes, and preserving and restoring natural barriers. To effectively mitigate tsunami risk, a comprehensive strategy is necessary. The effect of Murray Ridge on tsunami propagation is quantified, and our findings point out that in the presence of Murray Ridge, the wave arrives late and is amplified over the ridge and thereby amplified at the coast. In another study, we find that the Lakshadweep Islands serve as a wave barrier for the south-western coast of India and are shielding it. In yet another study, we observe that shelf width

DOI: 10.1201/9781003784500-7

plays a vital role in the arrival of tsunami waves along the west coast of India. The tsunamis from Andaman-Sumatra, too, have been analyzed, and we find a similar situation for tsunamis traveling over the Ninety East Ridge, where the tsunami's focus is towards the southern direction and the tsunami amplification is seen over the ridge, as in the case of Murray Ridge. The shielding effect of Sri Lanka on the southern tip of India, too, is quantified. All these studies emphasize the importance of selecting an appropriate resolution based on the specific requirements of the analysis.

The next critical challenge is reducing the destruction of man-made structures due to tsunamis, as discussed in Chapter 4. Safeguarding coastal infrastructure is vital to protecting both livelihoods and economic stability. Deep-sea structures generally withstand tsunami forces more effectively, since their design already accounts for hydrostatic and non-breaking wave pressures. Nevertheless, mapping the spatial distribution of maximum non-breaking wave pressure provides invaluable guidance for positioning future deep-sea structures. Closer to shore, however, breaking and broken wave forces dominate—and these can be far more severe. For example, the 2004 Indian Ocean tsunami devastated India's east coast—particularly the Tamil Nadu coast—and even well-prepared regions like Puducherry on the east coast of India suffered extensive losses. To comprehensively assess the vulnerability of Puducherry's coastal zone, it is therefore necessary to simulate various tsunami scenarios. By combining pressure-distribution plots for both non-breaking and breaking waves with region-specific bathymetry and coastal morphology, engineers can identify the most resilient locations for critical infrastructure—and ultimately mitigate the economic and social fallout of future tsunami events.

Even if a meteorite impact-generated tsunami is an infrequent event, the impact due to this is beyond anything expected and this is discussed in Chapter 5. In the past, an event called the Chicxulub impact, which occurred in the Yucatan Peninsula, was considered to be the reason for the extinction of many species, including dinosaurs. Hence, a deeper analysis will help identify locations for building safe houses for such catastrophic events. Although the occurrence of a meteorite impact-generated tsunami has a very low likelihood, if such an event ever occurs, it changes the face of any vulnerability studies done in the past.

A landslide tsunami occurs when a large amount of material, such as rocks or sediments, slides into the water along the coast, triggering a large wave that can cause widespread destruction and devastation. Tsunamis generated by the landslide are limited compared to those generated by submarine earthquakes and arise from slope instabilities. Chapter 6 explores the analysis of the landslide tsunami scenario, providing insights into potential inundation extents under various tsunami events. Understanding the range of possible impacts and planning for different scenarios is critical. The findings align with similar studies emphasizing the need for variability in tsunami impacts and improving disaster preparedness. For the Indian coast, the possibility of a landslide tsunami in both the Indus fan in the Arabian Sea and the Bengal fan in the Bay of Bengal cannot be ruled out. The Bengal fan sits on a vast reserve of gas hydrants and its dislocation could cause a landslide tsunami. With this in mind, a possible landslide tsunami was modeled using the COMCOT software, and the tsunami arrival times and their wave heights were computed at various locations.

7.2 FUTURE RESEARCH DIRECTIONS

There is a critical need to enhance tsunami safety measures and comprehensive disaster management strategies. Research should focus on further refining the models and methodologies used in these chapters. Enhancing the spatial resolution of satellite data, incorporating real-time monitoring systems, and integrating them can provide more accurate and comprehensive assessments of tsunami impacts as we all know that all three stages of a tsunami from generation to inundation depend on the modeling of the initial deformation at the source, as this is used as an initial condition to model the remaining two stages, i.e., propagation and run-up/inundation. Some available conventional codes neglect the dynamic ocean bottom. In the case of tsunamis generated by undersea earthquakes, these conventional techniques use static ocean bottom displacement, which doesn't give the exact upliftment of the seafloor near the epicenter. But in reality, the tsunamigenic earthquake focus is under the semi-half space/viscoelastic media, i.e., elastic solid layer and highly incompressible liquid layer, and also these classical codes neglect the weak coupling between the viscoelastic material, and also the rupture velocity and rise time in the process of projecting the ocean bottom displacement onto the seafloor. So, there is a need to develop tsunami modeling codes using the dynamic ocean bottom deformation and the above-mentioned limitations of the conventional codes.

To develop such codes, we must derive analytical solutions assuming different initial and boundary conditions. Inundation modeling needs high-resolution bathymetry and topography data of the study region, and the toposheets and NHO charts from different government agencies and institutions will be digitized for these studies. The development of high-resolution DEMs of the study regions using GIS and other relevant software is necessary. Preparation is required of travel time maps, inundation maps, and run-up maps for the coastal cities vulnerable to future tsunamis from earthquakes, landslides, volcanic eruptions, and meteorite impacts. These travel time maps, run-up maps, and inundation maps can be used by the local authorities for decision-making. Additionally, exploring the socio-economic dimensions of tsunami resilience, such as the role of governance and community participation, can offer valuable insights for building more resilient coastal communities.

Integrating multiple data sources such as GEBCO bathymetry, SRTM DEM, Landsat satellite imagery, and socioeconomic data enables a comprehensive assessment of tsunami impacts in coastal regions. The high-resolution GEBCO bathymetry data is critical for accurate tsunami propagation modeling, allowing for detailed analysis of inundation and run-up heights. Similarly, the SRTM DEM data with a high resolution provides essential information on elevation and slope, which are key factors influencing tsunami run-up and inundation patterns. Accurately predicting tsunami arrival times is crucial for early warning and evacuation. For the 2004 Sumatra event, the TUNAMI N2 model's predictions closely align with observed arrival times, highlighting its effectiveness in providing timely warnings. Accurate arrival time predictions minimize casualties and enable efficient evacuation and response efforts.

Index

G

H

I

K

L

M

For Product Safety Concerns and Information please contact our
EU representative GPSR@taylorandfrancis.com Taylor & Francis
Verlag GmbH, Kaufingerstraße 24, 80331 München, Germany